DECONTAMINATION
METHODOLOGIES
AND APPROACHES

The following States are Members of the International Atomic Energy Agency:

AFGHANISTAN	GAMBIA	NORWAY
ALBANIA	GEORGIA	OMAN
ALGERIA	GERMANY	PAKISTAN
ANGOLA	GHANA	PALAU
ANTIGUA AND BARBUDA	GREECE	PANAMA
ARGENTINA	GRENADA	PAPUA NEW GUINEA
ARMENIA	GUATEMALA	PARAGUAY
AUSTRALIA	GUINEA	PERU
AUSTRIA	GUYANA	PHILIPPINES
AZERBAIJAN	HAITI	POLAND
BAHAMAS	HOLY SEE	PORTUGAL
BAHRAIN	HONDURAS	QATAR
BANGLADESH	HUNGARY	REPUBLIC OF MOLDOVA
BARBADOS	ICELAND	ROMANIA
BELARUS	INDIA	RUSSIAN FEDERATION
BELGIUM	INDONESIA	RWANDA
BELIZE	IRAN, ISLAMIC REPUBLIC OF	SAINT KITTS AND NEVIS
BENIN	IRAQ	SAINT LUCIA
BOLIVIA, PLURINATIONAL	IRELAND	SAINT VINCENT AND
STATE OF	ISRAEL	THE GRENADINES
BOSNIA AND HERZEGOVINA	ITALY	SAMOA
BOTSWANA	JAMAICA	SAN MARINO
BRAZIL	JAPAN	SAUDI ARABIA
BRUNEI DARUSSALAM	JORDAN	SENEGAL
BULGARIA	KAZAKHSTAN	SERBIA
BURKINA FASO	KENYA	SEYCHELLES
BURUNDI	KOREA, REPUBLIC OF	SIERRA LEONE
CABO VERDE	KUWAIT	SINGAPORE
CAMBODIA	KYRGYZSTAN	SLOVAKIA
CAMEROON	LAO PEOPLE'S DEMOCRATIC	SLOVENIA
CANADA	REPUBLIC	SOUTH AFRICA
CENTRAL AFRICAN	LATVIA	SPAIN
REPUBLIC	LEBANON	SRI LANKA
CHAD	LESOTHO	SUDAN
CHILE	LIBERIA	SWEDEN
CHINA	LIBYA	SWITZERLAND
COLOMBIA	LIECHTENSTEIN	SYRIAN ARAB REPUBLIC
COMOROS	LITHUANIA	TAJIKISTAN
CONGO	LUXEMBOURG	THAILAND
COSTA RICA	MADAGASCAR	TOGO
CÔTE D'IVOIRE	MALAWI	TONGA
CROATIA	MALAYSIA	TRINIDAD AND TOBAGO
CUBA	MALI	TUNISIA
CYPRUS	MALTA	TÜRKİYE
CZECH REPUBLIC	MARSHALL ISLANDS	TURKMENISTAN
DEMOCRATIC REPUBLIC	MAURITANIA	UGANDA
OF THE CONGO	MAURITIUS	UKRAINE
DENMARK	MEXICO	UNITED ARAB EMIRATES
DJIBOUTI	MONACO	UNITED KINGDOM OF
DOMINICA	MONGOLIA	GREAT BRITAIN AND
DOMINICAN REPUBLIC	MONTENEGRO	NORTHERN IRELAND
ECUADOR	MOROCCO	UNITED REPUBLIC OF TANZANIA
EGYPT	MOZAMBIQUE	UNITED STATES OF AMERICA
EL SALVADOR	MYANMAR	URUGUAY
ERITREA	NAMIBIA	UZBEKISTAN
ESTONIA	NEPAL	VANUATU
ESWATINI	NETHERLANDS	VENEZUELA, BOLIVARIAN
ETHIOPIA	NEW ZEALAND	REPUBLIC OF
FIJI	NICARAGUA	VIET NAM
FINLAND	NIGER	YEMEN
FRANCE	NIGERIA	ZAMBIA
GABON	NORTH MACEDONIA	ZIMBABWE

The Agency's Statute was approved on 23 October 1956 by the Conference on the Statute of the IAEA held at United Nations Headquarters, New York; it entered into force on 29 July 1957. The Headquarters of the Agency are situated in Vienna. Its principal objective is "to accelerate and enlarge the contribution of atomic energy to peace, health and prosperity throughout the world".

IAEA
International Atomic Energy Agency

ORDERING LOCALLY

IAEA priced publications may be purchased from the sources listed below or from major local booksellers.

Orders for unpriced publications should be made directly to the IAEA. The contact details are given at the end of this list.

NORTH AMERICA

Bernan / Rowman & Littlefield

15250 NBN Way, Blue Ridge Summit, PA 17214, USA

Telephone: +1 800 462 6420 • Fax: +1 800 338 4550

Email: orders@rowman.com • Web site: www.rowman.com/bernan

REST OF WORLD

Please contact your preferred local supplier, or our lead distributor:

Eurospan Group

Gray's Inn House
127 Clerkenwell Road
London EC1R 5DB
United Kingdom

Trade orders and enquiries:

Telephone: +44 (0)176 760 4972 • Fax: +44 (0)176 760 1640
Email: eurospan@turpin-distribution.com

Individual orders:

www.eurospanbookstore.com/iaea

For further information:

Telephone: +44 (0)207 240 0856 • Fax: +44 (0)207 379 0609
Email: info@eurospangroup.com • Web site: www.eurospangroup.com

Orders for both priced and unpriced publications may be addressed directly to:

Marketing and Sales Unit
International Atomic Energy Agency
Vienna International Centre, PO Box 100, 1400 Vienna, Austria
Telephone: +43 1 2600 22529 or 22530 • Fax: +43 1 26007 22529
Email: sales.publications@iaea.org • Web site: www.iaea.org/publications

II. REPRODUCTIVE ORGANS

6. SUBFERTILITY AND INFERTILITY IN THE MALE A PERSISTENT DILEMMA

I.S. GOTTESMAN and J. BAIN

About 10% or more of all couples find themselves confronted by the problem of infertility (Mac Naughton, 1973), which is said to exist if pregnancy has not been achieved after one year of unprotected intercourse. Pure female factors account for approximately one-third of human infertility, pure male factors for one-third and a combination of male and female factors for the remainder. Therefore the male factor is prominent in the etiology of infertility. However, our understanding of the various facets of male subfertility remains rudimentary. Despite the advances in the biochemical, immunologic and morphologic parameters of human semen and in our knowledge of male reproductive physiology, we know little about basic pathophysiological events of male infertility, and hence we have not been able to develop highly successful therapeutic regimes for human oligospermia or azoospermia.

In the investigation of infertility the focus is not on one individual but rather on the couple. In the male we continue to explore those parameters that are readily accessible to investigation and try to make a definitive diagnosis. In this way we can proceed to develop a rational approach to therapy.

1. HISTORY TAKING

The importance of the history cannot be overemphasized. The history includes a general functional enquiry to rule out systemic disease but more specifically, endocrine dysfunction is sought after in great detail. It is important to know the marital status and whether there have been previous marriages or whether the patient has been responsible for pregnancies in the past. Of specific interest is a history of disease, drug ingestion, smoking habits and alcohol use (Engel and Morgan, 1973). Sy-

philis, gonorrhea, urethritis, prostatitis and orchitis, mumps-related or not, are important historical details (Söltz-Szöts, 1973).

One must ascertain whether sexual development has been normal and whether there has been any change in secondary sexual characteristics. The frequency, timing and mechanics of intercourse must be questioned to determine whether intravaginal ejaculation occurs at the appropriate time. Cryptorchidism and subsequent treatment as wll as genitourinary surgery may play a role. A familial or inherited disorder may be present and therefore one must inquire about a family history of infertility. Prior investigation and the outcome of previous treatment may serve as a guide for further investigation and therapy.

2. PHYSICAL EXAMINATION

The physical examination should be as complete as for any other general medical assessment with special emphasis on the endocrine and genitourinary systems. Height, weight and body proportions should be ascertained and secondary sexual characteristics carefully reviewed. Gynecomastia needs to be looked for and must be differentiated from increased adipose tissue. The breast examination is not complete without noting the presence or absence of galactorrhea. The penis and especially the urethral opening must be examined, first for its position and secondly to check for discharge and other signs of infection. The scrotum, testes, epididymis and vas are then examined with careful recording of testicular size and consistency. The testicular size of most men is 4 x 2 cm or greater, occupying a volume of 15 cm^3 or more, but there is wide variation even among fertile males. The presence of a varicocele should be determined by examining the patient in a standing position while he is performing the Valsalva maneuver. Also, other scrotal masses are felt for, such as a hydrocele, spermatocele or tumor.

The prostate gland is examined by digital palpation. The size and consistency of the prostate are assessed, nodules are felt for and an attempt is made to palpate the seminal vesicles. These follow the wall of the rectal canal on the same horizontal plane as the upper part of the prostate. The other hand is used to press over the bladder to try to squeeze the vesicles between the fingers. Normal vesicles are too soft to be distinguished from the bladder and therefore cannot be felt. After prostatic massage, the urethral fluid can be collected for bacteriologic assessment.

Miskin and Bain (1978) studied a group of men with B-mode ultrasound, measuring the width of the testis and correlating this to the sperm count. In a group of 11 men, each of whom had average sperm counts greater than 50 million/ml of ejaculate, the mean echographic width of the right testis was 2.25 cm and the left was 2.2 cm. As the sperm count decreased, the width of the testes was found to decrease, so that with an average sperm count of less than 10 million/ml, the mean widths were 1.58 cm on the right and 1.52 cm on the left. When testicular width categories are considered alone the sperm counts of all individuals within these categories can be seen to rise with increasing testicular width. Thus it appears that the size of the testes bears a direct relationship to the number of sperm produced. This is not surprising since the seminiferous tubules within which spermatogenesis occurs, comprise the major component of the testes.

Miskin and Bain (1978) also correlated the width of the testes by echography with that found on clinical examination. They found clinical measurements to be slightly greater than those by ultrasound and felt this was accounted for by skin and subcutaneous tissue that were measured clinically but were able to be delineated by ultrasound.

3. SEMEN ANALYSIS

In our laboratory, we request the freshest possible specimen to ensure an accurate assessment of sperm motility. Freund (1962) and Freund and Wiederman (1966) demonstrated that at room temperature (20-25° C) the percent motility with forward progression declines by 50% in the first seven hours after collection. They also showed that the sperm collection has to be at room temperature, because cooling to 41° F (5° C) results in a marked decrease in spermatozoal motility and forward progression. The rate of forward progression of 20 specimens at room temperature was compared with rates at body temperature (37°C) by Janich and MacLeod (1970). There was a marked but variable increase in the rate of forward progression in all 20 specimens when the temperature was raised from 23 to 25 to 37° C. We ask our patients to produce the specimen via masturbation into a sterile plastic bottle. Certain plastic containers may inhibit sperm motility since water soluble toxic substances may leak out of the plastic. This has been shown with a wide variety of plastic and rubber disposable syringes (Inchiosa, 1965) and more recently for plastic blood-bags (Guess et al., 1967; Jaeger and Rubin, 1970). That this actually leads to impaired spermatozoal motility and survival remains to be clearly demonstrated. We have demonstrated in our laboratory that with 100 consecutive semen analyses, the mean sperm motility with plastic collection bottles was 48% and with glass was 42%.

Because of significant variations in sperm count, morphology and motility in different specimens produced by the same man, it is essential to have two or more samples before further investigation or treatment is undertaken.

The specimen is produced after three days of sexual abstinence. Eliasson (1971, 1975) has advocated a set period of continence prior to analysis for the sake of standardization and to have seminal fluid which has not recently been depleted of sperm. With a standard period of abstinence the specimen is no longer a random sample of the patient's spermatozoal output at his usual frequency of intercourse (Freund, 1968). Individuals respond quite differently to experimental variations in frequency of intercourse and duration of continence with some males showing very large differences in spermatozoal output and some showing quite modes changes (Freund, 1962, 1963).

The specimen should not be produced by coitus interruptus since this may result in the loss of the first sperm-rich portion of the ejaculate. Condom collection leads to sperm death, immotility, inaccurate counts and volume measurements. If masturbation cannot be performed a special seminal pouch made of material that is not antagonistic to the sperm may be used for collection during intercourse.

Seminal fluid should be assessed for volume, viscosity, sperm motility, morphology and number. The presence of an increased number of white blood cells may suggest subclinical prostatitis. Semen volume is normally between 1 and 5 ml. The percentage of motile sperm will vary among laboratories depending on the criteria that are used in considering what is purposeul forward progression. The definition of normospermia is in constant flux. In the past infertility was said to exist if the count was less than 60 million/ml (MacLeod and Hotchkiss, 1946; Sanders and Macomber, 1929; Tyler, 1951; Tyler and Singher, 1956). Amelar (1966) and Williams (1964) (independently) felt that oligospermia was less than 40 million/ml and then in 1965, MacLeod stated that oligospermia should be said to exist at sperm counts less than 20 million/ml, a revision of his earlier opinion (MacLeod, 1965b; MacLeod and Hotchkiss, 1946).

MacLeod (1973) even questioned this figure and suggested 10 million/ml as the cut-off point for oligospermia. The conception rate rose if the sperm count exceeded 10 million/ml, but decreased to 50% if the count fell below this level (MacLeod, 1973; Van Zyl et al., 1975). A drastic drop in conception rate (less than 20%) occurred if the sperm count was less than 5 million/ml. The majority of authors currently consider that oligospermia begins with sperm counts less than 20 million/ml, with a distinct drop in fertility occurring below this level (Freund and Peterson, 1976).

There are no firm data indicating what percentage of sperm must have normal morphology to ensure fertility, but there is the suggestion that this must be greater than 50-60% (Heller et al., 1950).

The presence of a large number of leukocytes (greater than 20/hpf) suggests the possibility of a genitourinary tract infection. If this exists, the semen should be cultured for bacteria as well as for T-strain mycoplasma.

A major factor influencing the number of spermatozoa in each specimen is the number of sperma-tozoa that are available for ejaculation in the extratesticular reserve (i.e. sperm that are present in the vasa deferentia and epididymides). About two thirds of the sperm in any given specimen come from this anatomical region (Freund, 1962, 1963). This epididymal reserve is one of the important factors in determining the number of sperm in the ejaculate and provides the physiological basis for the effect of the period of continence prior to collection of the sample. The loss of sperm through previous ejaculations is the chief factor affecting epididymal sperm reserves and therefore sperm count.

4. FACTORS AFFECTING MALE FERTILITY

Table 1 demonstrates factors affecting male fertility from three different centers. The greatest incidence of infection in the male genitournary tract (26%) is found in males whose sperm counts are less than 10,000,000/ml. These infections include prostatovesiculitis (28%). gonorrhea (7%), syphilis (2%), and mumps orchitis (3%) (Van Zyl et al., 1976). Prostatovesiculitis, which may result in azoospermia, can esily be missed yet is a potentially reversible cause of oligospermia. Eliasson (1975) has described a case of azoospermia caused by prostatovesiculitis in a patient whose fertility was restored after treatment for 18 months.

The incidence of varicocele is higher in males with low sperm counts. The incidence of varicocele in males with counts of 10-20 million/ml has been reported to be 34% and with counts less than 20

Table 1. Etiological factors in male infertility.

Factors	Dubin and Amelar (1971)*	Van Zyl et al. (1975)*	Gottesman and Bain** (unpubl. results)
Infections	—	26	9.3
Varicocele	39	24	13.3
Chromosomal anomalies	3	12	1.1***
Cryptorchidism	4	3	3.9
Endocrine factors	9	2	—
Obstructive azoospermia	7	3	0.24
	n = 1294	n = 596	n = 822

* Males with sperm counts less than 10 million/ml.
** Males with sperm counts less than 20 million/ml.
*** All Klinefelter's syndrome and all azoospermic.

millon/ml, 33%. The incidence of the other subgroups with sperm counts greater than 20 million vary between 16 and 27% (MacLeod and Gold, 1951; MacLeod, 1965a; Van Zyl et al., 1975, 1976; Dubin and Amelar, 1971).

Men with large varicoceles whose sperm counts greater than 60 million/ml have impregnated their female partner (Van Zyl et al., 1976) Men with small varicoceles may have more significant abnormalities of their semen parameters, than men with larger varicoceles. The varicocele affects semen morphology and motility to a greater extent than it affects sperm count, which may be normal or low (MacLeod, 1965a).

Chromosomal anomalies, especially those relating to the sex chromosomes, may have a severe and permanent effect on all semen parameters. Kjessler (1973) found the highest incidence of chromosomal abnormalities to occur among males with sperm counts of less than 10 million/ml. The sperm counts, though, in two patients were found to be greater than 60 million/ml. These were two autosomal abnormalities with 46XYq and 45XYt (DqDq) whose counts were 150 million/ml and 69 million/ml respectively. Paulsen (1974) emphasizes that spermatogenesis may be normal, minimally impaired or severely impaired in those males with Klinefelter's or Klinefelter's mosaic syndromes.

Cryptorchidism was not found in males with sperm counts greater than 20 million/ml, whereas the incidence with a count less than 20 million/ml was 3% (Van Zyl et al., 1975).

Among Van Zyl's (1976) patients, only 2% had confirmed systemic endocrine disturbances. One patient with diabetes had already fathered two children; two patients had a chromphobe adenoma and were oligospermic; three men had thyroid disorders and there was one patient with hypothalamic obesity.

Other factors that have been noted to affect semen analyses have been stress, nutrition, trauma, allergic factors, retrograde ejaculation, and the use of marijuana (Van Zyl et al., 1976).

We have accumulated data on 822 males who were referred for a fertility investigation. These men were divided into several sperm count categories, as seen in Table 2.

Men with specific pathological entities were then grouped according to sperm count category. Those

Table 2. Number and percentage of men in various sperm count categories.

Count (million/ml)	Number of men	Percentage of total number investigated
> 0	101	12.3
> 0- 5	109	13.3
> 5-20	144	17.5
>20-40	135	16.4
>40	333	40.5

who gave a history of having had orchitis are listed in Table 3.

It can be seen that only a small number of males who presented for infertility investigation had a history of orchitis (25 out of 822 or 3% of the total). What is striking though is that of these 25 men, 20 or 80% were oligospermic and 14 or 56% severely so with counts less than 5 million/ml. Thus it would appear that even though orchitis is an infrequent findings, when a history of this is obtained there will be a very high incidence of oligospermia.

Other studies have shown that prostatovesiculitis has its greatest incidence in those males with oligospermia. In our study the incidence of infection in the oligospermic group was 26 out of 822 or 3.16% of the total. For the normospermic groups it was 18 out of 822 or 2.7% of the total. Although the difference may not be significant, it is still important to diagnose genitourinary tract infection because of the possibility of reversing the infertility by specific treatment (Table 4).

The presence of a varicocele may be responsible for decreased sperm counts as well as abnormal morphology and motility. Our results indicate that there was no significant difference in the occurrence of a varicocele in the different sperm count or motility categories except when sperm count ex-

Table 3. Men with orchitis divided according to sperm count category.

Count (million/ml)	Number in each group	Percentage within each group
0	7/101	6.9
> 0- 5	7/109	6.4
> 5-20	6/144	4.2
>20-40	3/135	2.2
>40	2/333	0.6

IAEA NUCLEAR ENERGY SERIES No. NW-T-1.38

DECONTAMINATION METHODOLOGIES AND APPROACHES

INTERNATIONAL ATOMIC ENERGY AGENCY
VIENNA, 2023

COPYRIGHT NOTICE

© IAEA, 2023

Printed by the IAEA in Austria
November 2023
STI/PUB/2066

IAEA Library Cataloguing in Publication Data

Names: International Atomic Energy Agency.
Title: Decontamination methodologies and approaches / International Atomic Energy Agency.
Description: Vienna : International Atomic Energy Agency, 2023. | Series: Nuclear Energy Series, ISSN 1995-7807 ; no. NW-T-1.38 | Includes bibliographical references.
Identifiers: IAEAL 23-01614 | ISBN 978–92–0–144123–2 (paperback : alk. paper) | ISBN 978–92–0–143923–9 (pdf) | ISBN 978–92–0–144023–5 (epub)
Subjects: LCSH: Radioactive decontamination. | Radioactive decontamination — Methodology. | Radioactive decontamination — Safety measures
Classification: UDC 621.039.59 | STI/PUB/2066

FOREWORD

The IAEA's statutory role is to "seek to accelerate and enlarge the contribution of atomic energy to peace, health and prosperity throughout the world". Among other functions, the IAEA is authorized to "foster the exchange of scientific and technical information on peaceful uses of atomic energy". One way this is achieved is through a range of technical publications including the IAEA Nuclear Energy Series.

The IAEA Nuclear Energy Series comprises publications designed to further the use of nuclear technologies in support of sustainable development, to advance nuclear science and technology, catalyse innovation and build capacity to support the existing and expanded use of nuclear power and nuclear science applications. The publications include information covering all policy, technological and management aspects of the definition and implementation of activities involving the peaceful use of nuclear technology. While the guidance provided in IAEA Nuclear Energy Series publications does not constitute Member States' consensus, it has undergone internal peer review and been made available to Member States for comment prior to publication.

The IAEA safety standards establish fundamental principles, requirements and recommendations to ensure nuclear safety and serve as a global reference for protecting people and the environment from harmful effects of ionizing radiation.

When IAEA Nuclear Energy Series publications address safety, it is ensured that the IAEA safety standards are referred to as the current boundary conditions for the application of nuclear technology.

Each IAEA Nuclear Energy Series publication on predisposal management of radioactive waste is structured in a way that guides the user through the development, implementation and evaluation of a successful predisposal management programme. It is suggested that all these publications be consulted when developing the application specific processes.

Detailed theoretical background information is limited in favour of a basic overview and understanding of the relevant thematic area. Extensive references are provided in each publication for users wishing to obtain additional technical details about various subjects within the scope of the publication.

This publication was developed by international experts and advisors from several Member States. The IAEA wishes to acknowledge the valuable assistance provided by the contributors and reviewers listed at the end of the publication. The IAEA officer responsible for this publication was W. Meyer of the Division of Nuclear Fuel Cycle and Waste Technology.

CONTENTS

1. INTRODUCTION

1.1. BACKGROUND

Decontamination procedures have been routinely adopted from other industries for the nuclear industry since the 1960s. In the 1970s, decontamination methods and techniques became more complex, and an integrated assessment of technical performance, environmental factors and costs became the norm for decontamination projects. Currently, decontamination is a topic of great interest as the need for it is increasing because of the growing number of redundant facilities facing decommissioning in the near term, the performance of upgrading activities, alterations or repairs, and the inspection and maintenance required to allow continuation of cost effective and safe operation of existing installations. The decontamination methods used will influence the further treatment, conditioning and even free release of material generated during this process.

Large decontamination projects will often require a combination of decontamination processes, and it is necessary to review major assumptions regularly during implementation. Therefore, the selection of the best strategy and technology during the planning process is a crucial responsibility for operating organizations.

The IAEA has addressed decontamination in several publications. However, the existing guidance publications are outdated, and in some cases obsolete, while newer publications provide information about technologies, but much less guidance about their selection. This publication provides information on the selection of a decontamination approach and the decision making process.

1.2. OBJECTIVE

The objective of this publication is to provide information on factors (technological, safety, environmental, organizational) relevant to decontamination strategies and methods. It also presents an overview of the general approaches to the holistic assessment of relevant factors leading to the selection of a preferred method or methods, as well as examples of experience and lessons learned from the impacts of those factors on the success (or failure) of specific decontamination projects.

Target groups for this IAEA publication include operating organizations and especially their decontamination managers, waste management organizations, research and development institutes and universities active in this realm, supply chain companies, and consultants/contractors to operators and regulatory bodies. Given the significant role decontamination plays in nuclear decommissioning, it is expected that all those involved in planning and implementation of decommissioning will find this publication of interest. Guidance and recommendations provided here in relation to identified good practices represent experts' opinions but are not made on the basis of a consensus of all Member States.

1.3. SCOPE

This publication focuses on decision making in the planning of decontamination projects or campaigns for nuclear and radiological facilities. It includes consideration of relevant factors and how individual factors are weighed, merged and combined for the purposes of an integrated decision on the preferred approach. The publication also addresses decontamination following accidental releases. Cleanup of mining and milling operations or contaminated soils is outside its scope. Similarly, the publication does not deal with personal decontamination resulting from inadvertent exposure to local contamination (i.e. skin contamination or inhalation of airborne contamination).

This publication does not provide technical details on individual decontamination methods and technologies. Detailed information is available on the IAEA's Wiki[1] pages and the Deactivation and Decommissioning Knowledge Management Information Tool (D&D KM-IT)[2], managed by Florida International University, United States of America (USA).

1.4. STRUCTURE

Following the introductory section, Section 2 expands on management considerations when selecting a decontamination methodology. Section 3 summarizes the available decontamination technologies. Section 4 provides an analysis of factors relevant to the selection of decontamination methodologies. Section 5 suggests approaches for selecting a decontamination methodology. Section 6 provides stepwise, simplified coverage of activities to be considered during the implementation of decontamination. Special decontamination cases in response to incidents and accidents are given in Section 7, and Section 8 provides conclusions. Supplemental information is provided in a series of annexes, which present relevant case studies and experience concerning specific aspects of decontamination.

2. MANAGEMENT CONSIDERATIONS WHEN SELECTING A DECONTAMINATION METHODOLOGY

The establishment and implementation of a national programme for radioactive waste management, including decontamination methodologies, are influenced by many factors, ranging from the legislation in the State to the views of the stakeholders (industry and public) of the decontamination project. Addressing these factors and incorporating them into the final decontamination proposal will prevent or reduce challenges during the implementation phase. This section introduces the various management factors influencing the selection of a decontamination methodology.

2.1. SETTING OBJECTIVES FOR A DECONTAMINATION PROGRAMME

According to the IAEA Nuclear Safety and Security Glossary 2022 (Interim) Edition [1], decontamination is "The complete or partial removal of *contamination* by a deliberate physical, chemical or biological *process*." It is important to be aware that:

— Decontamination is not the elimination of radioactivity, just its transfer to a different location and/ or form;
— Decontamination is not an objective per se: it serves the purposes of maintenance, inspection, repair, or dismantling of equipment or components as part of decommissioning;
— There is no single decontamination strategy or technique that is suitable for all forms of contamination.

Before planning decontamination, it is essential to define the objectives and a realistic end point of the process. General objectives of a decontamination may include, among others:

---

[1] See https://nucleus.iaea.org/sites/connect/IDNpublic/Pages/IDN-Wiki-Introduction.aspx.
[2] See https://www.dndkm.org.

(a)   Reduction of radiation source terms during operation and consequently anticipated occupational exposures in preparation for large scale maintenance, refurbishment or decommissioning. Radiation and/or contaminations levels are often too high to permit such activities or to continue plant operation (frequent crew changes are required to comply with individual exposure limits, and many qualified workers are either unavailable or too costly to consider). This objective can be achieved by:

   (i)   Reducing radiation fields to allow personnel access, increase the duration of planned activities, or reduce requirements for shielding;

   (ii)   Reducing surface contamination to reduce requirements for respiratory protection (especially relevant to alpha contamination).

(b)   Reduction of the volume of equipment and materials requiring disposal in licensed repositories by downgrading the waste category, for example from low level waste to exempt waste, which is eligible for unrestricted release, or to waste bound for specified restricted release (nuclear or non-nuclear).

(c)   Salvage of equipment and materials for intact reuse or recycle.

(d)   Enablement of better and optimized use of financial, human and technical resources.

Establishing concise objectives and goals and communicating these to all stakeholders is the cornerstone of a successful decontamination management programme. Objectives, based on the national policies and strategies for radioactive waste management, will be influenced by regulatory, economic and stakeholder drivers, together with other strategic priorities. It has to be kept in mind that for all these processes local, regional, national, international and site specific standards or requirements also have to be considered. When established and agreed, the objectives will provide a framework and information for the selection of the decontamination method(s). The objectives and programme may be refined or adjusted during the process, as more detailed information on the effluent characteristics, process performance and secondary waste management options becomes available.

2.2. ALIGNMENT WITH NATIONAL POLICIES AND STRATEGIES

States are advised to have a national policy and strategy for managing their radioactive waste in place (Fig. 1), including decontamination technologies. The following definitions are taken from Policies and Strategies for Radioactive Waste Management (IAEA Nuclear Energy Series No. NW-G-1.1) [2], which needs to be consulted for a more in-depth discussion of policy and strategy.

"**Policy** is a set of established goals or requirements for the safe management of spent fuel and radioactive waste; it normally defines national roles and responsibilities. As such, policy is mainly established by the national government; policy may also be codified in the national legislative system.

"**Strategy** is the means for achieving the goals and requirements set out in the national policy for the safe management of spent fuel and radioactive waste. Strategy is normally established by the relevant waste owner or operator, either a governmental agency or a private entity. The national policy may be elaborated in several different strategies. The individual strategies may address different types of waste (e.g. reactor waste, decommissioning waste, institutional waste, etc.) or waste belonging to different owners."

Policies and strategies need to address the overall life cycle management of the waste. Choices in one area (e.g. preferred disposal method) will affect the options available in other areas (e.g. required treatments and performance of the conditioned waste form).

Defined policies and strategies need to be flexible in order to accommodate future changes, such as new national circumstances (e.g. legislative changes and plans for new nuclear facilities), new international agreements and/or new advances in technologies. These changing circumstances may lead to the modification of current practices, the discontinuation of past practices, or even the introduction of new

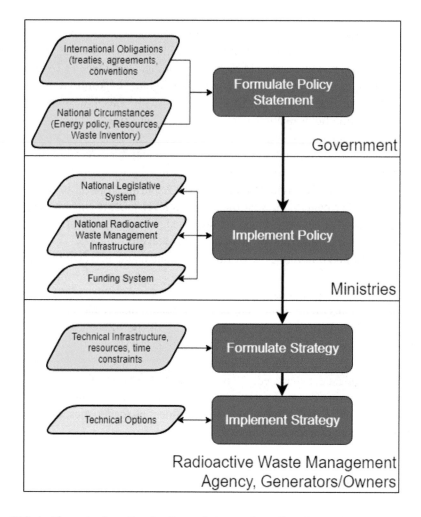

FIG. 1. Elements of a national policy and strategy for radioactive waste management.

ones. The lead in making changes needs to be taken by the body responsible for the initial formulation of the policy (typically the government) and strategy, but all relevant parties in the State need to be involved in the process.

2.3. STAKEHOLDER ENGAGEMENT

Developing and implementing any step in radioactive waste management (including decontamination) involves many entities, among them subject matter experts, designers, builders, regulatory bodies, operators and managers. Because of this complex network, a change in one of the steps may influence other programmes or resources, or regulatory compliance. It is important to identify and conduct discussions with parties that will be affected by the development and implementation of the individual steps. The essential point here is acceptance by the industry and public to process material released for unrestricted use from a nuclear facility.

The stakeholder engagement strategy and its implementation require careful review and evaluation. Developments in a proposed new decontamination process need to be documented and communicated clearly in a timely manner to affected parties.

The public represents a key stakeholder group, and therefore their acceptance of important steps in waste management is very important. In many cases, the existing nuclear facilities already have well established mechanisms for public information, consultation and education in place, and the same mechanisms could be used to explain new decontamination processes and facilities. In several States, public consultation is a legislative requirement and can be also reflected in the radioactive waste policy.

2.4. SAFETY ASPECTS

Safety requirements and guidance for radioactive waste management in general are set out and discussed in detail in a number of IAEA publications, including Safety Standards Series publications and Safety Reports. These publications range from basic principles of radioactive waste management safety to detailed safety requirements for various scenarios. For decontamination, safety requirements and guidance can be found in the following publications:

(a) Leadership and Management for Safety (IAEA Safety Standards Series No. GSR Part 2) [3];
(b) Safety Assessment for Facilities and Activities (IAEA Safety Standards Series No. GSR Part 4 (Rev. 1)) [4];
(c) Predisposal Management of Radioactive Waste (IAEA Safety Standards Series No. GSR Part 5) [5];
(d) Predisposal Management of Radioactive Waste from Nuclear Power Plants and Research Reactors (IAEA Safety Standards Series No. SSG-40) [6];
(e) Predisposal Management of Radioactive Waste from Nuclear Fuel Cycle Facilities (IAEA Safety Standards Series No. SSG-41) [7];
(f) Storage of Radioactive Waste (IAEA Safety Standards Series No. WS-G-6.1) [8];
(g) Application of the Concepts of Exclusion, Exemption and Clearance (IAEA Safety Standards Series No. RS-G-1.7) [9];
(h) Leadership, Management and Culture for Safety in Radioactive Waste Management (IAEA Safety Standards Series No. GSG-16) [10];
(i) Regulatory Control of Radioactive Discharges to the Environment (IAEA Safety Standards Series No. GSG-9) [11];
(j) Classification of Radioactive Waste (IAEA Safety Standards Series No. GSG-1) [12];
(k) The Safety Case and Safety Assessment for the Predisposal Management of Radioactive Waste (IAEA Safety Standards Series No. GSG-3) [13];
(l) Decommissioning of Nuclear Power Plants, Research Reactors and Other Nuclear Fuel Cycle Facilities (IAEA Safety Standards Series No. SSG-47) [14];
(m) Decommissioning of Facilities (IAEA Safety Standards Series No. GSR Part 6) [15].

Safety aspects need to be fully considered in the selection, design, licensing, construction and operation of a decontamination system for radioactive waste treatment. These considerations are documented in a number of previous IAEA publications [16–22] and publications from other organizations [23, 24] that have been published over the years and deal with various technical, managerial and safety related aspects of decontamination activities.

2.5. CONFORMANCE TO WASTE ACCEPTANCE CRITERIA

All waste management facilities (for processing, storage and disposal) ideally have acceptance requirements, conditions or guidelines. The waste acceptance criteria (WAC) are based on the safety case and licence for that specific facility.

Secondary waste is usually generated as a result of decontamination activities. When selecting a decontamination process, it is preferable to use decontamination processes that generate the smallest possible volume of secondary waste and from which the waste generated fulfils the WAC of the on-site treatment facilities. The availability of on-site waste treatment facilities with the ability to deal with all decontamination waste is an important factor in selecting the decontamination process.

2.6. WASTE CLASSIFICATION AND CATEGORIZATION

2.6.1. Waste classification

The IAEA waste classification system, based on the minimum disposal requirements for different classes of radioactive waste, is described in GSG-1 [12]. It is based on a hierarchy of disposal requirements for waste packages containing higher activity levels and radionuclides with longer half-lives, necessitating higher degrees of containment and isolation. The suggested disposal routes indicate the minimum acceptable degree of containment and isolation. National policies or legislation typically establish the ways in which waste streams are handled and/or classified (Fig. 2), based on their physical characteristics. For the selection of a decontamination technology, it is important to consider the reduction in activity levels of materials such as steel and concrete for reuse or recycling purposes.

2.6.2. Waste categorization

The basic categories of waste are described as 'unconditioned' waste and 'conditioned' waste, with the latter meaning waste that is ready for disposal. Classifying these wastes based solely on radioactivity and radionuclide content is not viable for all waste types during every step in the waste management process. Categorization of wastes to include factors such as origin, physical state, type of waste, properties and process options also provides the basis for a consistent approach for their management. Each major waste category can further be subdivided based on:

(a) Point of origin — source of the raw waste;
(b) Physical state — liquid, gaseous or solid;
(c) Type — dry solids, resin, sludges, slurry, metal, combustible and compactable;
(d) Properties — radiological, physical, chemical (in some cases, biological) and volume.

These properties are used to select the applicable technical options for decontamination for a given end point. In some cases, non-radiological properties (e.g. chemical toxicity) can influence the selection of the decontamination process [25].

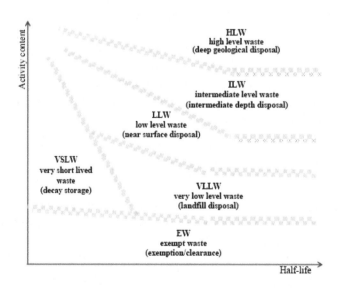

FIG. 2. Waste classification based on the quantities and qualities of the material. The specific thresholds for each type of waste are typically set out in national policies or legislation.

2.7. ENVIRONMENTAL, HEALTH AND SAFETY PROGRAMME

Every decontamination project will require a dedicated occupational environmental, health and safety (EHS) programme, which is developed by the operating organization or its contractors. This is essential to limit the occupational radiation exposure time of the workers and needs to be compatible with the operating organization's overall site specific EHS protection programme. The optimization of protection and safety is typically subject to the ALARA (as low as reasonably achievable) approach [1]. The programme will include the identification and control of anticipated hazards and the monitoring of performance. The programme will also address individual EHS protection, including training for decontamination contractor and site services personnel.

Discharges of non-radioactive and radioactive solutions during decontamination processes are likely to be important and subject to regulatory interest, consumption of resources (power and chemical reagents) and greenhouse gas emissions, while noise and visual impact may also be of concern. Thus, many Member States require an environmental impact assessment (EIA) before regulatory acceptance of new waste management facilities or processes, or significant changes to existing processes. The EIA takes place before irreversible decisions are made regarding a decontamination project and covers all stages of the project from technology selection and preparation through construction and operation. The EIA needs to prove that the environment is protected throughout the various decontamination operations.

2.8. QUALITY ASSURANCE AND CONTROL

Quality assurance and quality control are important aspects of any decontamination programme, and provide assurance that the waste management function has been performed according to procedure and that it meets the specified requirements as set forth in the policy or regulations. This includes documented demonstration that (1) the decontamination method selected is suitable for the set objectives, (2) the decontamination tools and substances are satisfactorily designed and manufactured and (3) the personnel in charge are qualified and trained.

Quality assurance — a set of records associated with a facility or process that verifies management of that facility or process according to a structured system that considers the appropriate safety, technical and non-technical requirements.

Quality control — a demonstration of the acceptability of the product (i.e. each finished, conditioned waste package) compared to its specified requirements. Quality control is based on inspections, sampling and testing to measure the process output and ensure that it is within the defined acceptable ranges.

Applicable national standards (such as those issued by ASTM, the American Society of Mechanical Engineers (ASME), the Deutsches Institut für Normung (DIN), Association Française de Normalisation (AFNOR), GOST, the CSA Group and the British Standards Institution (BSI)) and international standards (such as those issued by the International Organization for Standardization (ISO) 9000/9001/14000 series and the European Committee for Standardization (CEN)) could be considered when developing the quality assurance and quality control processes for a waste conditioning programme.

2.9. ECONOMIC FACTORS

The economics of dismantling is a complex subject that not only involves technological choices, but also local factors such as the availability of a suitable supply chain and the cost of skilled resources, specialized materials, equipment and infrastructure. Several different cost elements need to be considered in selecting a decontamination process, including:

— Research and development;
— Engineering and capital costs;

— Selection of technology;
— Licensing;
— Construction;
— Operation and maintenance;
— Managing of secondary waste;
— Decommissioning;
— Infrastructure development.

The last two items (decommissioning of decontamination equipment and related infrastructure development) are often overlooked when accounting for costs related to the selection of a decontamination programme. The exact infrastructure needs will depend on the waste types, the chosen technology and the desired location.

Cost effectiveness needs to be balanced against many of the factors already discussed in this section, such as regulatory compliance, system complexity and throughput.

2.10. LICENSING A DECONTAMINATION CAMPAIGN

The extent of the licensing process for a decontamination campaign will generally depend on whether the process is a routinely approved activity (i.e. the regular cleanout of accessible areas) or a new activity. The licensing process can take a variable length of time, depending on technical and managerial aspects of regulatory interest, and therefore can be relevant to the selection of the preferred decontamination approach.

The documentation needed for licensing of the decontamination process includes the description (including sequence of activities), analysis and safety assessment regarding possible changes in the radiological state and risk situations. Additional documents could be specific to the management of secondary waste, the assessment of decontaminated structures, systems and components (SSCs) and their capability to perform any required functions. In general, the documents submitted to the regulator(s) include the following:

— Safety analysis, including details such as description of the initial state and radiological characterization of the affected SSCs; the overall decontamination project, the scope of each phase of the decontamination, the activities that might entail modifications to the safety conditions; safety assessment, including safety and radiation protection reference regulations, the identification of hazards in normal and accident conditions, and measures to mitigate risks.
— Technical specifications for safety related plant components and systems that are to be kept in operation or on standby during decontamination.
— Radiological monitoring and surveillance programmes.
— Industrial safety analysis to consider hazards, for example falls from heights, stored energy and confirmed spaces.
— Descriptions of the organization, roles and responsibilities of plant personnel and contractors, including certifications, qualifications and training arrangements.
— Quality management manual and assurance of consistency between the quality management programme of the operating organization and that of contractors performing decontamination.
— Radiological protection manual, including the organization, regulations and procedures for radiation protection.
— On-site emergency plan specifying the organization and provisions in place to prevent incidents and accidents, and to mitigate their consequences if they occur.
— Plan for the management of all radioactive waste resulting from decontamination.

Depending on national legislation and practices, the regulatory bodies review and approve operating procedures for the decontamination; alternatively, these procedures may be reviewed by site inspectors without the need for formal approval.

In accordance with Fundamental Safety Principles (IAEA Safety Standards Series No. SF-1) [26], the prime responsibility for a decontamination project rests with the licensee or operating organization, and includes aspects such as:

— Preparation of a detailed decontamination plan and procedures;
— Conducting safety assessments;
— Ensuring availability of adequate technical, human and financial resources;
— Establishment of a quality management programme;
— Ensuring the health and safety of the personnel, the public and the environment;
— Minimizing the generation of radioactive waste;
— Ensuring safety and security at all stages of waste generation and management during decontamination;
— Acquiring and preserving records relevant to decontamination.

2.11. APPLICATION OF REMOTE OR MOBILE DECONTAMINATION

Remote technology is generally limited to applications where the prevailing conditions (high radiation fields) make manual operations prohibitive. Remote technologies are increasingly used in decontamination processes with the following objectives:

(a) Reduce radiation exposure to workers;
(b) Permit faster processing and throughput;
(c) Increase reliability;
(d) Enhance safety.

As the tasks become more complex, the functions of, and the inputs to, the remote system increase, and remote operations become impractical or prohibitively costly [27]. Given the high costs of remotely operated equipment and the difficulties inherent to retrieving a robot from an area with high levels of radiation, reliability is a key factor in remote decontamination. Therefore, the design of remotely operated equipment needs to include provisions such as redundancy, maintenance and easy decontamination after operation.

Many companies offer mobile decontamination trailers for nuclear decontamination purposes. Mobile services allow the operating organizations to deploy decontamination processes with little capital expenditure and reduced set-up time. The equipment can be managed by vendor personnel who are trained in the decontamination services offered, the operation of related equipment and good health physics practices. The drawback is that only small items can be decontaminated using a mobile trailer.

2.12. NO DECONTAMINATION AS THE OPTIMAL APPROACH

Decontamination needs to be implemented if significant benefits can be achieved. There are several factors mitigating against decontamination activities. Firstly, decontamination may involve radiation exposure of the personnel. Decontamination needs to be justified in terms of the reduction of dose in subsequent plant operations, care and maintenance or dismantling activities. Similarly, the financial costs of decontamination need to be justified in terms of the savings that will accrue from subsequent activities such as surveillance and maintenance, dismantling and waste disposal.

From a financial point of view, justification for decontamination (e.g. releasing some steelwork to unrestricted release level) needs to consider the cost required to achieve release (including treatment and disposal of secondary waste) compared with both the scrap value of the material and the cost of disposal as radioactive waste that will be avoided. The impacts from possible incidents during the decontamination activities (e.g. spreading of contamination to adjacent areas) need to be taken into account. Having considered all the factors, it could be concluded that no decontamination might be the best approach under the prevailing circumstances.

2.13. RADIATION PROTECTION

Although radiation exposure during decontamination is generally low and the benefit substantial, the potential radiation exposure needs to be considered during the selection process. Part of the selection process could consider the economic, social or other benefits in relation to the health detriment radiation exposure may cause. Responsibility for evaluating radiation exposure risk usually falls on national radiation protection authorities, although these authorities are likely to need input so that a fully informed decision can ultimately be made. This can be reviewed if new information becomes available.

3. DECONTAMINATION TECHNIQUES

This section presents a general summary of available decontamination techniques, without providing unnecessary details on individual decontamination methods and technologies. More detailed information is available elsewhere [14–25, 27]). This section does not attempt to be exhaustive; it is intended as a first orientation for users.

3.1. INTRODUCTION

Decontamination, as a step in the pretreatment of solid waste, is defined as the complete or partial removal of contamination by a deliberate physical, chemical or biological process, to reduce radiological hazards or transform wastes from a higher category to a lower one. The objectives of decontamination are to:

— Reduce radiation exposure;
— Allow contaminated equipment to be reused;
— Reduce the volume of equipment and materials requiring storage and disposal in licensed facilities for storage and disposal of radioactive waste;
— Restore the site and facility, or parts thereof, to an unconditional use state;
— Remove loose radioactive contaminants;
— Reduce the residual radioactive inventory in a protective storage mode (for public health and safety reasons) or shorten the protective storage period.

There is no universal decontamination technique that is suitable for all applications. The selection of technologies depends on the decontamination activity needed and the outcome planned. Considerations include:

— The material to be decontaminated and the state of its surface (chemistry of oxide layers, roughness, irregularities, geometry);
— The process effectiveness required;
— The operability/simplicity/reliability required;

— The licensing requirements (ease of environmental and safety compliance, compliance with existing permits and/or licences and safety publications);
— The physical conditions and other industrial aspects of the environment (temperature, humidity, layout, access and egress, available space for all decontamination related activities);
— The post-decontamination requirements (reuse, WAC of the treatment facility, or free release requirements);
— Occupational exposure;
— Industrial safety (material storage and handling, leakage, ingestion of toxic gases);
— The management of primary and secondary waste.

The main types of decontamination techniques (schematic presentation in Fig. 3) can be subdivided into four categories:

(a) Chemical methods are very common and can vary between destructive and non- destructive techniques. The chemicals used and their application method can be tailored to the requirements of the process.
(b) Electrochemical methods generally involve the application of an electric field across a surface in order to remove contamination. An example is electropolishing, a technique used in a non-nuclear environment to give metallic surfaces a polished finish.
(c) Energetic methods are innovative and involve supplying energy in the form of heat, electromagnetic radiation or sound in order to remove contamination from a surface.

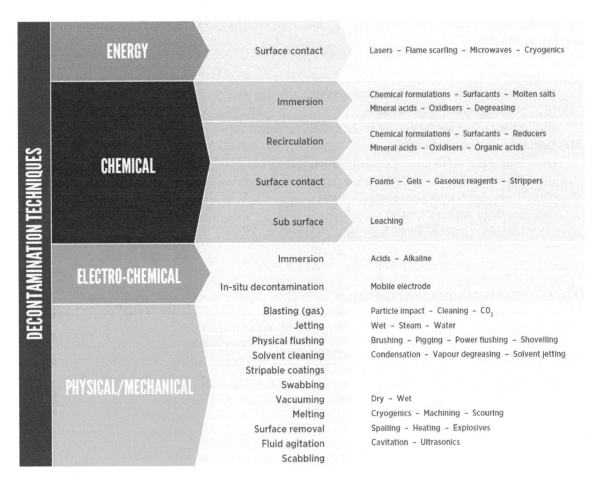

FIG. 3. Schematic presentation of decontamination techniques. Courtesy of O. Thompson, National Nuclear Laboratory Ltd, United Kingdom.

(d) Physical and mechanical techniques are by far the most varied and well established in terms of technical maturity. In general, such methods involve the removal of contamination through various physical means.

3.2. LEVEL OF DECONTAMINATION REQUIRED (DECONTAMINATION FACTOR)

The decontamination factor (DF) is the quantitative factor describing the decontamination efficiency. It is used in many Member States as a method to compare decontamination technologies. According to the definition given in the IAEA Safety Glossary [1], DF is the ratio of the activity per unit area (or per unit mass or volume) before a particular decontamination technique is applied to the activity per unit area (or per unit mass or volume) after application of the technique:

$$DF = \frac{\text{activity measured before treatment}}{\text{activity measured after treatment}}$$

The effective removal of contamination by a decontamination process is a very important criterion for the selection of a process. DF is a parameter used to measure the process effectiveness. The relation between DF and the fraction of the removed activity is specified in Table 1.

The desirable value of DF in a decontamination project depends on the specific circumstances. In some cases, relatively low DFs (2–3) can be acceptable, but in many cases DFs of the order of 10 are required. It is important that the DF value is not viewed as the only criterion for the selection of a decontamination technique. This is because DF values are strongly dependent on the operating conditions and surface conditions, such as:

(a) System related:
    (i) Material type (carbon steel, stainless steel, nonferrous);
    (ii) Smooth versus rough surfaces;
    (iii) Speed of the decontamination chemicals (turbulent versus laminar flow).
(b) Contamination characteristics:
    (i) Extent and uniformity of contamination spread;
    (ii) Physical, chemical and radiological nature of contamination layers;
    (iii) Fixed versus non-fixed contamination;
    (iv) Single or multiple layers;
    (v) Soluble versus insoluble deposits.

TABLE 1. DECONTAMINATION FACTORS

DF	% of activity removed $(1 - 1/DF) \times 100\%$
2	50%
5	80%
10	90%
50	98%
100	99%

(c) Corrosion effects:
  (i) Corrosion type on base metal;
  (ii) Absorption of activity on corrosion layer.
(d) Solvent stability and concentration used:
  (i) Temperature of solvent;
  (ii) Concentration of solvent;
  (iii) Chemistry control during the process.

Often the pursuit of higher DFs to reduce a dose rate to an acceptable level, prolonged application times or repeated application of chemicals can increase the volume of secondary waste.

3.3. ENERGY DECONTAMINATION TECHNIQUES

Energy decontamination methods (Table 2) involve contacting the contaminant with a high energy source for the contaminant to be evaporated. The type of energy source used depends upon the composition of the base material, the nature of the contaminant to be removed and the planned future use of the material or item.

Laser decontamination (ablation) is achieved by directing the output of a high power laser through optics to the surface of the component, preferably with a motion control system. The surface of the component absorbs the laser energy rapidly and as a result evaporates instantaneously, removing contamination without the use of any cleaning agents (Fig. 4). Laser selection is based on the following cutting properties: vaporize, melt and blow, melt blow and burn, thermal stress cracking, scribing, cold cutting and burning stabilized laser cutting. Melt and blow or fusion cutting uses high pressure gas to blow molten material away from the cutting area, greatly decreasing the power requirement.

The advantages and disadvantages of using energy decontamination methods are presented in Table 3.

TABLE 2. APPLICABILITY OF ENERGY DECONTAMINATION TECHNIQUES

Techniques	Applicability of use			Decontamination factor (DF)	Cost
	Large volume/ closed systems	Segmented parts	Building and surface structures		
Lasers	x	x	x	Medium → high	High
Microwaves		x	x	Medium → high	Medium
Flame scarfing	x	x	x	Medium → high	High
Cryogenic		x		Medium → high	High

TABLE 3. ADVANTAGES AND DISADVANTAGES OF USING ENERGY DECONTAMINATION METHODS

Advantages	Disadvantages
— Can be very effective even for complex geometries and internal surfaces — Can be applied for processing equipment without dismantling (i.e. complex objects such as primary circuits, heat exchangers, extraction columns, pipes and valves in situ) — Good DFs can be achieved — Very reliable operation — High performance — Remote application	— DFs are not as good on porous surfaces — Can be relatively expensive — Closed systems are needed to prevent airborne contamination

FIG. 4. Laser ablation of metal. Courtesy of Energiewerke Nord GmbH, Germany.

3.4. CHEMICAL DECONTAMINATION TECHNIQUES

Chemical decontamination methods involve contacting the contaminant with a chemical reagent to elicit a reaction. The selection of the type of reagent used depends upon the nature of the contaminant to be removed, the planned future use of the material or item and the ability of the user to deal with any secondary waste generated, including the spent decontamination solution (see Table 4).

Figure 5 representing chemical decontamination is an example of an aqueous process; it is usually only applied to remove contamination that is in the upper layers of a surface (from a few micrometres to a few hundred micrometres only).

There are four main ways a chemical decontamination agent can be applied:

(a) Immersion — degreasing agents, mineral acids, surfactants, molten salts — chemical formulations are placed within a bath or tank and the contaminated material is fully submerged.

(b) Recirculation — redox agents, mineral acids, surfactants — chemical formulations are used in a system that is regenerated so as to minimize the quantity of reagent used.

(c) Surface contact — foams, gels, strippable coatings, gaseous reagents and mists/fogs, which are applied directly to a contaminated surface.

(d) Subsurface — leaching agents are applied to porous materials to loosen and/or remove contamination at depth. These agents tend to be proprietary and are not routinely used in the nuclear industry.

The advantages and disadvantages of using chemical decontamination agents are presented in Table 5.

TABLE 4. APPLICABILITY OF CHEMICAL DECONTAMINATION TECHNIQUES

Technique	Technology	Applicability of use			Decontamination factor (DF)	Cost
		Large volume/ closed systems	Segmented parts	Building and surface structures		
Chemical immersion	Chemical formulations	x	x	x	Medium to high	Medium
	Degreasing agents		x			
	Mineral acids		x			
	Surfactants		x			
	Molten salts		x			
	REDOX agents		x			
	Organic acids		x			
	Leaching agents					
Surface contact	Foam	x		x	Medium to high	Medium
	Chemical gels	x	x	x		
	Chemical pastes	x	x			
	Chemical fog	x	x			
	Gaseous reagents	x				
	Stripping agents		x	x		
Subsurface contact	Leaching agents			x	Medium	Medium
Microbial degradation	Addition of acidic secreting organisms	x		x	Low to medium	Low to medium

FIG. 5. Example of a chemical decontamination of metals systems. Courtesy of Necsa, South Africa.

TABLE 5. ADVANTAGES AND DISADVANTAGES OF USING CHEMICAL DECONTAMINATION AGENTS

Advantages	Disadvantages
— Can be very effective even for complex geometries and internal surfaces — Contamination with complex composition of radionuclides in oxide forms (U_xO_y and Pu_xO_y) can be treated — Can be relatively inexpensive, especially in cases where additional equipment or personnel are not required — Can be applied for processing equipment without dismantling (i.e. complex objects such as primary circuits, heat exchangers, extraction columns, pipes and valves in situ) — Established method in the nuclear industry — Depending on the combination of chemicals used, a good DF can be achieved — Closed systems can prevent airborne contamination — Can be applied in a variety of different ways to optimize results achieved (misting, spraying, pastes, gels)	— Requires effluent treatment/disposal of a relatively high volume of secondary waste (i.e. the aggressive and/or hazardous solutions (acids/bases with complex compositions) produced) — Hard to reach surfaces are difficult to treat and it may not be possible to ascertain the effectiveness — To aid effectiveness, the solution may need to be heated (typically 70–90°C) — High risk of corrosion associated with the process, and production of secondary wastes, which may impact on plant infrastructure (drains) — DF is not as good on porous surfaces Some of the most effective chemicals used are not compatible with downstream options

3.5. ELECTROCHEMICAL DECONTAMINATION TECHNIQUES

In principle, electrochemical decontamination (Table 6) may be considered as the application of an electrical field to chemical decontamination agents to assist in their effectiveness. It may be considered the opposite of electroplating, as metal layers are removed from a surface rather than added as a coating. Electrochemical decontamination is usually applied by immersion of the contaminated item in an electrolyte bath or by passing a pad over the surface to be decontaminated, as indicated in Figs 6 and 7.

The electric current causes anodic dissolution and the removal of metal and oxide layers from the component. The electrolyte is regenerated continuously by recirculation. Electrochemical techniques can only be applied to conducting surfaces. However, they have high success rates and often yield high decontamination factors. They can be applied in two ways:

TABLE 6. APPLICABILITY OF ELECTROCHEMICAL DECONTAMINATION TECHNIQUES

Technique	Technology	Applicability of use			Decontamination factor (DF)	Cost
		Large volume/ closed systems	Segmented parts	Building and surface structures		
Immersion solutions	Acidic/ alkaline electrolytes	x	x		High	High
In situ	Mobile brush		x		High	Medium

FIG. 6. Electropolishing of a highly contaminated stainless steel piece in the KRB, a boiling water reactor in Grundremmingen, Germany. Courtesy of M. Laraia.

FIG. 7. Removing contamination embedded inside piping. Courtesy of South African Nuclear Energy Corporation.

(a) Immersion — a contaminated object is submerged in a chemical formulation. An electrical current is then applied to enhance the chemical decontamination process.

(b) In situ — mobile electrodes can be directly applied to a contaminated surface.

The advantages and disadvantages of using electrochemical decontamination are presented in Table 7.

TABLE 7. ADVANTAGES AND DISADVANTAGES OF USING ELECTROCHEMICAL DECONTAMINATION

Advantages	Disadvantages
— Can achieve greater DFs than chemical techniques alone — Low volumes of secondary waste — Good for contamination hot spots	— Requires an energy source — Cannot be used on complicated geometries — Requires skilled operators

3.6. PHYSICAL AND MECHANICAL DECONTAMINATION TECHNIQUES

Physical and mechanical methods of decontamination are based on the destruction and subsequent removal of a contaminated surface without distinction of the chemical and/or physical form of the contaminant (see Table 8).

TABLE 8. APPLICABILITY OF MECHANICAL DECONTAMINATION TECHNIQUES

Technique	Technology	Applicability of use			Decontamination factor (DF)	Cost
		Large volume/ closed systems	Segmented parts	Building and surface structure		
Blasting	Ice		x	x	Low to high	Medium
	Metal		x	x		
	Soft abrasive		x	x		
	Grit		x	x		
	Ceramics		x	x		
Jetting	Steam		x	x	Low to high	Low
	Water	x	x	x		
Flushing	Brushing				Low to high	Medium
	Pigging			x		
	Power flushing	x				
	Shovelling			x		
Solvent cleaning	Condensation		x		Medium to high	Medium
	Vapour degreasing		x			
Strippable coatings		x	x	x	Medium	Low

TABLE 8. APPLICABILITY OF MECHANICAL DECONTAMINATION TECHNIQUES (cont.)

Technique	Technology	Applicability of use			Decontamination factor (DF)	Cost
		Large volume/ closed systems	Segmented parts	Building and surface structure		
Swabbing			x	x	Low to medium	Low
Vacuuming			x	x	Low	Low
Melting			x		Medium to high	High
Surface removal	Cryogenic		x	x	Medium to high	Medium to high
	Machining			x		
	Scouring			x		
	Spalling			x		
	Heating			x		
	Explosives			x		
	Vibrocleaning			x		
Fluid agitation (cavitation)	Hydrosonic/ ultrasonic		x	x	Low to medium	Medium

These mechanical decontamination systems are generally considered to be abrasive techniques, as they act to remove the surface layer of contaminated substrates to separate the contamination from the object, lowering the classification status (see Figs 8 and 9). The removed contamination can then be collected, treated if necessary, and disposed of as waste. However, as during this process the volume of waste is greatly reduced, it will also mean that the management costs are lower, especially the associated disposal costs.

In the nuclear industry, mechanical decontamination is the most common decontamination technique, and as such a wide variety of methods are available at high maturity levels. The advantages and disadvantages of using physical and mechanical decontamination techniques are presented in Table 9.

The main consideration, however, is the expected effectiveness of the process. Based on these considerations, one or more decontamination processes can be selected as likely candidates.

FIG. 8. Example of surface decontamination using steam cleaning. Courtesy of South African Nuclear Energy Corporation.

(a) Before

(b) After

FIG. 9. Item before and after grit blasting. Courtesy of Cyclife Sweden.

4. FACTORS RELEVANT TO THE SELECTION OF A DECONTAMINATION METHODOLOGY

4.1. INTRODUCTION

The contamination of components is the result of various physical and physicochemical processes, and the quantity of contaminants on surfaces depends on many factors, such as the type of the base

TABLE 9. ADVANTAGES AND DISADVANTAGES OF USING PHYSICAL AND MECHANICAL DECONTAMINATION TECHNIQUES

Advantages	Disadvantages
— Can be used on a wide range of surfaces — Easily available — Solid waste can be collected easily using vacuums or sweepers — Can be an automated process or used robotically to reduce worker exposure to high doses — Easy application — High level of maturity and much nuclear expertise — Absence of secondary wastes in liquid form reduces treatment costs	— Secondary wastes are in the form of airborne contamination, debris, dust, rust and scale, slag and, for wet processes, discharged liquid waste (slurries) — Formation of airborne contamination requires additional protection and filtration — The contaminated surface needs to be accessible — Requires a power supply — Requires staff to deploy the technology

material, the surface roughness, the degree of corrosion and the surface or material porosity, as well as the physicochemical properties of the fluid, such as pressure, temperature and pH value. Before selecting a suitable decontamination process, a detailed analysis of the contamination and possible decontamination process options is necessary, for instance:

— Location of the contamination (inner versus outer surfaces of closed fluid systems);
— Material (metal, concrete, plastic, wood);
— History of operation (determine contamination profile);
— Composition of the contamination (oxide, crud, sludge);
— Distribution of contamination (surface, cracks, homogeneous distribution in bulk material);
— Possible exposure to workers and the public;
— Management based on exposure level (recycling versus disposal);
— Quantity and type of secondary waste arising from decontamination;
— Available waste management routes for processing, storage and/or disposal;
— Availability of suitably qualified and experienced workers;
— Time and cost.

A decontamination programme also requires that a facility be capable of treating arising secondary waste (processing chemical solutions, aerosols, debris), as this waste will contain concentrated activities.

The amount of contamination to be removed influences the selection process. The objective could be a reduction of the radiation dose rate of the component, the removal of loose or semi-loose contamination, the decategorization of the component, or even the free release of the treated material.

4.2. TECHNICAL INFORMATION REQUIRED

The technical performance of decontamination techniques depends on various parameters, including the composition of base material to be decontaminated, the physical presence of radioactive contamination (mobile or fixed, depth of penetration), the contamination characteristics, the maturity of the technology, the level of decontamination required (decontamination factor) and the potential hazards of the current contamination situation. These parameters are discussed in more detail and need to be considered when evaluating the different decontamination methods.

4.2.1. Composition of base material to be decontaminated

No single (universal) decontamination process is suitable for use on all types of materials when considering chemical decontamination. For example, stainless steel, carbon steel, aluminium and concrete will all interact differently with the chemicals proposed, and thus the base material composition influences the selection of possible decontamination chemicals. For instance, the decontamination rate could be reduced by solutions penetrating into the pore structure of material, or solutions that corrode component surface, generating additional corrosion products that needs to be removed. Therefore, a thorough analysis of the contaminated target material (type, pore structure, corrosion) needs to be undertaken when selecting the most suitable decontamination process.

4.2.2. Physical presence of contamination

The physical presence (structure) of the activity earmarked for decontamination is one of the most important criteria for the selection of the decontamination process.

4.2.2.1. Non-fixed (mobile) contamination

Mobile contamination includes activity of loose particles (dust) deposited on stagnant areas of surfaces as well as activity (Fe_2O_3) loosely bound onto the surface itself. These forms of contamination can be removed by selecting decontamination techniques such as wiping, brushing, vacuuming or washing using appropriate liquid chemicals. Removal of dust needs to be managed carefully to prevent airborne resuspension of dust containing activity.

4.2.2.2. Fixed (adherent) contamination

This form of contamination (activity) is chemically or physically bound to the material surface and is normally found as corrosion product in a surface corrosion layer. As fixed contamination can accumulate in specific areas susceptible to corrosion, high concentrations can increase radiation exposure levels. Selection of the decontamination technique for this situation will be based on the removal of the corrosion layer and damaged base metal could be passivated by electropolishing to reduce its chemical activity. Removal of fixed contamination by polishing and grinding needs to be performed in designed facilities to minimize spreading of volatile reaction products. Fixed contamination may exude, leach, or otherwise become non-fixed due to the release of the corrosion layer.

4.2.2.3. Penetration into cracks and crevices (pores)

This type of activity diffusing into surface pores of the material and over time resurfacing due to changes in temperature and humidity and oxidation reactions to the material surface (plutonium from stainless steel, uranium from porous cement) may require repeated applications of the selected decontamination technique.

Typical decontamination processes for porous contaminated surfaces (e.g. concrete) are based on layered removal of the external surfaces. Large concrete surfaces can be decontaminated either by scabbler, shaver, or steel grit blast. Applications with shavers resulting in a smooth decontaminated surface are preferred, as the concrete surface after using scabbling and steel grit blast technologies will contain cracks and pores that will be more prone to contamination than before. Layered removal of concrete surfaces needs to be managed carefully to prevent airborne contamination (cross-contamination) to other sections and personnel.

Activity in metal pores in components, such as valves, pumps, tanks and filter housings, can be partially decontaminated with the use of pore penetrating solutions such as foams, gels or strippable coatings. These polymeric solutions enter the pore structure, allowing the incorporation of radionuclides into the solution structure. Upon solidification, the formed coatings (with encapsulated activity) can be readily removed. The major disadvantage of this technique is that, due to low DFs, repeated applications are required.

4.2.3. Contamination characteristics

The selection of a decontamination process to remove contamination depends to a large extent upon understanding of the radiological, chemical and physical origin and the quantity of activity to be removed. Different chemical compositions (speciation) of activity interact differently to the chemicals proposed for decontamination and therefore the composition of the activity influences the selection of the decontamination chemicals. Understanding the interaction between the activity and decontamination process will permit a more straightforward decontamination process with a minimum amount of secondary waste. Examples of contamination characteristics to be determined are:

— The extent and uniformity of contamination spread;
— The type of radiation (alpha, beta, gamma);
— The physical, chemical and radiological nature of contamination layers;
— Fixed versus non-fixed contamination;
— Single or multiple layers;
— Soluble versus insoluble deposits.

Activity (contamination) containing fissile material is a potential criticality risk should radionuclide concentration occur during the decontamination treatment process. Appropriate criticality risk assessment and criticality prevention and management procedures will be required as part of the selection of the decontamination process.

4.2.4. Activity on degraded coated surfaces protecting base materials

Coatings such as paints and laminates and plastic or metallic lining are used to protect surfaces from contamination. As coatings degrade, their protective function is compromised, and decontamination could be problematic if activity has partially penetrated through the coating layer onto the base material.

To evaluate the contamination situation, the coating needs to be removed safely (minimizing airborne dust) to determine the composition and activity of the contamination as well as the status of the base material. Once removed, the contaminated base material can be characterized and the appropriated decontamination process selected.

4.2.5. Technology maturity

Ideally, a technological matured decontamination process can be considered. It needs to have proven operational effectiveness, be well documented and have been validated by other users. Additionally, it is important if the earmarked process has been demonstrated in projects with similar contamination and working conditions.

The ease with which chemicals and components for a decontamination process can be procured, installed and operated is an important factor in selecting a process. Ideally, the selected process can make use of readily available equipment and tools ('off the shelf') or decontamination equipment that is already in use in other projects, rather than needing to be designed and manufactured according to specifications.

Another key factor for the selection of a decontamination process is the reliability of the equipment used in the process. This is usually expressed as mean time between failures, which is the anticipated time between inherent failures of a decontamination system during operation. A shorter mean time between failures means that the system might be out of service and in need of repair more often.

4.2.6. Potential hazards

Potential radiological and non-radiological hazards have to be considered during the selection of a decontamination process. These hazards are briefly introduced in Sections 4.2.6.1–4.2.6.3. A thorough analysis of the potential hazards associated with the available decontamination processes needs to be performed, as a basis for selecting the most suitable one for the prevailing situation.

4.2.6.1. External exposures

During the preparatory phase, when equipment is being set up and system alterations are being made (e.g. the installation of filling and draining connections and component isolation), exposures occur as workers get close to the SSC that needs to be decontaminated. Selection of the decontamination technology could involve pre-decontamination dismantling to remove potential sources of radiation.

4.2.6.2. Internal exposures due to inhalation

During decontamination using hydroblasting, wet blasting, grinders and scabblers airborne contaminants such as contaminated droplets or dust can be generated. Appropriate decontamination techniques that limit internal exposures due to inhalation need to be selected. This is essential in the presence of significant concentrations of alpha emitters.

4.2.6.3. Industrial hazards

Potential industrial (non-radiological) hazards during the execution of a decontamination process could stem from the following:

— Storage of large volumes of hazardous solutions;
— Obstructions causing increased risks;
— Mechanical problems encountered during the decontamination campaign (failure of operational equipment; pressurized fluids leaking; skin exposure to or inhalation of toxic vapours);
— Secondary waste (generated process gases need to be managed to prevent fires or explosions, and abrasive blasting of aluminium can generate fine particles that may ignite, causing fire);
— General industrial hazards (failure of high pressure water pipes used for cutting; electrical shock or burns), eye hazards (due to dust, debris and noise).

5. APPROACH TO SELECTING A DECONTAMINATION METHODOLOGY

Selection of a decontamination methodology normally begins with the acquisition and assessment of data for all candidate processes and the consideration of all relevant factors inherent to each of them. A range of decontamination options is then formulated together with a preliminary decontamination plan for each option. Different selection methodologies are available, and this section provides an approach for selecting integrated decontamination technologies.

5.1. THE SELECTION PROCESS

The main selection process step is to identify which decontamination processes are available and suitable for the specific application. Technologies can be selected for a very specific purpose (for a single specific waste stream) or their flexibility to handle a range of contaminated material adequately. The selection process needs to be robust and transparent, with the outcomes suitably documented and justified. Some examples of key considerations before embarking on a selection process are:

— Understanding of the nature and properties of the object to be decontaminated (radiological/chemical/physical/biological properties);
— Definition of the desired performance, including goals and attributes or criteria that are important for the selection of the decontamination technology (e.g. reuse, free release);
— Acquisition of any further data that may be required to underpin the assessment (WAC of treatment or conditioning facilities to accept the decontaminated object);
— Performance of an options assessment and selection exercise comparing potential options against the predetermined management criteria or attributes (environmental assessments, economic factors, safety, regulatory restrictions, design requirements, operating requirements);
— Determination of the current performance (TRL readiness and volume reduction factor) for existing available technologies.

Figure 10 provides an overview of the selection process activities. Sections 5.1.1–5.1.5 provide further information on selected steps.

5.1.1. Step 1: evaluation of all relevant non-technical factors

The first step in a selection process includes the evaluation of all relevant non-technical factors, including the decontamination goals (reuse, free release), regulatory framework, safety requirements,

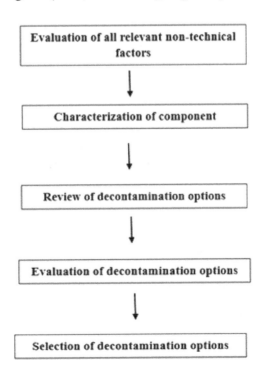

FIG. 10. Overview of the process for technology selection.

occupational and public radiation exposure, environmental protection, secondary waste management and operational requirements and limits. The first five of these non-technical factors are briefly described.

5.1.1.1. *Goals and attributes*

One of the key steps in defining the selection process for a new processing technology is the use of known strategic priorities to clearly define the required and desired goals and attributes, together with derived selection criteria for a proposed conditioning treatment system. This includes consideration of regional, national, international and site specific standards or requirements.

5.1.1.2. *Regulatory compliance (legal framework)*

Local, regional, national and international regulations or agreements (e.g. the 1992 Convention for the Protection of the Marine Environment of the North-East Atlantic (OSPAR Convention) [28]) are used to identify regulatory or otherwise authorized requirements (facility, processing, environmental, transport), including pending changes.

5.1.1.3. *Safety requirements and risks*

The overall safety risks and hazards that would be encountered in using a particular decontamination strategy need to be evaluated, considering installation, operation, maintenance, occupational risks from industrial hazards and decontamination activities.

5.1.1.4. *Occupational and public radiation exposure*

Specific regulatory and administrative limits for both occupational and public exposure need to be defined. Assessing routes to meet these limits include design walkdowns, installation, operation, maintenance and decommissioning of decontamination facilities. Public dose evaluations may require assessment of exposure from multiple pathways because of airborne, liquid and, in some instances, solid (highway transportation) emissions.

5.1.1.5. *Environmental protection*

Objectives related to environmental stewardship and meeting regulatory requirements on radioactive discharges and secondary wastes or other environmental protection goals may be appropriate. National, local and site specific discharge and/or disposal limits are related to:

— Activity, chemistry, and other parameters (e.g. concentrations, totals);
— Frequency;
— Mass, volume;
— Rate of discharge or disposal.

5.1.2. Step 2: characterization of components

The key information required before embarking on a selection process is:

(a) Understanding of the nature and properties of the object to be decontaminated (type, location, physical properties);
(b) Characterization of the nature and properties of the activity (contamination) on the object to be decontaminated (radiological, chemical, physical, dose rate).

5.1.3. Step 3: review of decontamination options

5.1.3.1. Operational effectiveness

Ideally, the decontamination process to be considered needs to have proven its operational effectiveness, be well documented and have been validated by other users. For a semi-quantitative appraisal of the decontamination processes for consideration, technology readiness levels (TRLs) could be used. These refer to an internationally recognized method of assessing technology maturity of elements of a programme during its acquisition. A technology readiness assessment (TRA) reviews programme concepts, technology requirements and validated/documented technology capabilities. TRLs are based on a scale from 1 to 9, with 1 being basic technological research and 9 being the most mature technology.

Definitions of TRLs are given by the European Commission [29] as follows:

"— TRL 1: basic principles observed
— TRL 2: technology concept formulated
— TRL 3: experimental proof of concept
— TRL 4: technology validated in lab
— TRL 5: technology validated in relevant environment (industrially relevant environment in the case of key enabling technologies)
— TRL 6: technology demonstrated in relevant environment (industrially relevant environment in the case of key enabling technologies)
— TRL 7: system prototype demonstration in operational environment
— TRL 8: system complete and qualified
— TRL 9: actual system proven in operational environment (competitive manufacturing in the case of key enabling technologies; or in space)".

The different decontamination technologies need to be compared against the requirements mentioned in step 1.

5.1.3.2. Technical goals required for a decontamination technology

Establishing technical goals requires a detailed understanding of a technology as well as the systems and/or technologies that it may interface with. Technical goals are typically related to specific design and performance attributes, including pressure bands, physical size, process rates, secondary waste generation rates, maintenance requirements and operability. When establishing the goals and attributes, it is also important to consider the possible effects from integrating a technology into a more complex system and/or connecting it to existing facility services (installation in an already radioactive plant location), or how it will be affected by this. Technical risks (e.g. new technologies or widely varying waste compositions) may be appropriately included here.

5.1.4. Step 4: evaluation of decontamination options (detailed technology screening)

Once the process goals and characteristics of the object to be decontaminated have been identified or defined, and possible technologies identified, selection of a specific technology can commence.

In this step, the range, capabilities and limitations of the relevant technologies need to be reviewed carefully to ensure that their attributes are properly understood prior to elimination or further consideration.

One approach to this step is to use a spreadsheet data capture technique and several rounds of evaluations to identify candidate technologies. The first round of evaluations focuses on eliminating technologies that are not viable when considered relative to goals and waste stream characteristics.

When a technology appears to be viable, a more detailed assessment of process configuration options will be performed, which could include the following elements:

— Whether there is a need for multiple technologies to produce the desired results;
— Compatibility of a technology with other existing or potential technologies or processes;
— Impact of hazard evaluation (radiological, industrial and chemical hazards) of the proposed process;
— Ease of attaining regulatory compliance or approvals for installation and operation;
— Size and weight of components;
— Proof of experience for similar system applications at other plants and availability of existing designs;
— Space requirements and availability;
— Logistics;
— Availability of support equipment needed for system set-up and material handling equipment during operation;
— General interface requirements, including availability, their location relative to the planned system location, the need for modifications, regulatory requirements for the type of interface required and the general capacity requirements (voltage, current, pressure);
— Maintenance requirements and ease of equipment access and/or removal and decontamination;
— Ease of decontamination relative to future disposal or transport of the system;
— Decommissioning and disposal of contaminated equipment, which need to be considered during the evaluation and design phase, as they have implications for the cost and waste volumes;
— Effect on secondary waste retention and processing capacity and capabilities, and storage or disposal.

5.1.5. Step 5: selection of decontamination options

Once the detailed technology screening process has been completed and the candidate technologies that meet the required constraints and attributes have been determined, process configuration options need to be developed. In some instances, a single technology may produce the desired results, while in others a single technology may not be a viable option, so a combination of techniques could be required. Similarly, several conditioning process options may be necessary, using permutations and combinations of the different techniques, each capable of meeting the objectives and targets but differing from one another in many respects.

Each process option needs to be developed in sufficient detail to allow it to be compared using the criteria or attributes. This includes the development of chemical flowsheets and consideration of technology-specific and external design considerations and constraints, as well as the need for remote operation and mobilization based on activity concentration and radiation fields (e.g. Fukushima post-accident processing), or other radiological considerations, such as shielding and containment.

In creating the process options, it also is important to ensure that they have a common starting point and a common end point. This allows the scope of the option selection process to be defined, and ensures that options are evaluated and compared using a reasonable standard basis.

Once the processing options have been developed, evaluations of each configuration against an attribute or criterion will most likely involve a combination of both qualitative and quantitative assessments. Many key attributes, such as costs and secondary waste volumes, can be quantified. Others, such as ease of operation or technical flexibility, may need more qualitative assessment and justification to be recorded.

5.2. INDUSTRIAL BENCHMARKING

Industrial benchmarking could be considered during the selection process, contacting current technology users, suppliers and facilities with similar waste streams. This approach often results in capturing very useful information, reducing the level of effort required for selection and significantly improving the potential for developing a successful decontamination management process. Benchmarking requires a careful review of both the data in question and the specifics of the processes to which those data apply. Variations in the reporting period (e.g. calendar year, fuel cycle), format, content, results, units and facility types, and operational differences will affect the results. Similar applications need to be benchmarked where practical.

6. KEY CONSIDERATIONS FOR IMPLEMENTING THE SELECTED OPTION

The implementation process for the selected decontamination method(s) is a series of sequential steps that refine a process from a relatively crude vision to a very detailed, operable system that can produce the desired results. The implementation scope and sequence of each step in the process may vary depending on the selected decontamination method; therefore, the information is intentionally not prescriptive, but rather structured as considerations to be used on a case by case basis.

Regardless of the project scope or implementation sequence, this section contains detail to ensure that all applicable actions are incorporated into the design, construction, operation and decommissioning phases of the treatment process. The key steps in the implementation process are the following:

(a) Organization (including production of procedures) and management;
(b) Integrated management systems;
(c) Environmental impact and safety case;
(d) Inactive commissioning, testing and demonstration;
(e) Commissioning;
(f) Operation;
(g) Demonstration of success and validation of processes;
(h) Report to the regulatory body and other stakeholders.

This section provides an overview of the implementation processes before the commissioning of a selected decontamination technology.

6.1. ORGANIZATION AND MANAGEMENT

A cost effective decontamination programme requires an organizational structure that has no ambiguities or overlaps in reporting and communication lines. Authority and responsibility levels and distribution of work need to be described in procedures. The entire decontamination operation needs to be supervised by a decontamination manager with general functions such as coordinating and supervising, advising on planning, selection and implementation of processes, and tackling unexpected problems. The decontamination manager leads a team reporting to him/her, and is likely to fulfil, among others, the following duties:

— Manage group training;
— Manage interactions with manufacturers and/or vendors of field instruments and tools, health physics, on-site and off-site laboratories;
— Assess, consent to, monitor and suggest how to improve decontamination procedures;
— Evaluate field data and trends, including decontamination effluents and secondary waste;
— Audit tools and materials that go in or come out of the work area;

— Supervise the radioactive waste segregation, storage, handling, transfer routes and doses;
— Ensure that materials to be used in decontamination are supplied regularly;
— Use managerial functions (leadership and team spirit) to create and maintain safety culture;
— Ensure that necessary data are suitably collected, checked, audited and processed to allow timely decisions to be made;
— Draft and submit reports for top management.

The decontamination manager needs to ensure that external contractors, who have specialist knowledge of the decontamination work itself, be trained in plant orientation and specific radiological protection requirements (e.g. exposure control and use of personal protective equipment (PPE)) associated with the area in which they are deployed.

6.2. INTEGRATED MANAGEMENT SYSTEM

Similar to safety requirements, management actions are defined in licence regulations and industrial standards that affect all, or portions of, the implementation process. If required, each phase of the implementation process, as outlined in this section, would have specific management system controls and criteria, such as:

(a) Design reviews: level of review, by what parties, design requirements, confirmation of compliance with applicable licence and other publications, and required reviews prior to approval;
(b) Trials: equipment, waste stream, monitoring and results evaluation process;
(c) Procurement: materials of construction, their origin and supplier certifications, verification of performance and test standards used during manufacture;
(d) Construction: worker qualification and testing, inspections including visual and other non-destructive methods, acceptance criteria, hold points for project or process review, and hydrostatic and other performance testing;
(e) Acceptance testing and commissioning: conformation with design and performance criteria, including operational, radiological and other criteria;
(f) Operation: operator training and qualification, maintenance activities and performance assessments;
(g) Decontamination operations: waste removal, handling, treatment of packaging, transport and disposal.

6.3. ENVIRONMENTAL IMPACT ASESSMENT AND SAFETY CASE

The legal requirements will require the operator to submit EIAs and a safety case to the regulatory body in support of an application for a licence or for authorization. The suite of documents supporting EIAs and a safety case need to include a series of assessment reports, commensurate with the complexity of the facility and the magnitude and likelihood of potential exposures and risks posed by the facility. They need to address, as a minimum, the following elements:

— Description of the decontamination facility and its components, equipment and systems;
— Site characterization;
— Organizational control of the operations;
— Procedures and operational manuals for activities with significant safety implications;
— Commissioning plans and schedules;
— Safety assessment;
— EIAs, where applicable;
— Monitoring programme;

— Training programme for staff;
— Nuclear material accountancy and safeguards aspects, where applicable;
— Arrangements for nuclear security and physical protection of radioactive material;
— Emergency preparedness and response plan;
— Integrated management system;
— Decommissioning of the decontamination facility.

6.4. INACTIVE COMMISSIONING, TESTING AND DEMONSTRATION

The successful development of a decontamination process, including testing of operating procedures for decontaminating large scale SSCs, requires comprehensive studies and laboratory and pilot plant tests. While testing is mandatory for innovative techniques, even a well proven technique requires some testing, based on conditions prevailing in the actual project.

Decontamination methods could initially be tested in the laboratory, using simulated or ideally real samples. However, laboratory conditions will inevitably be different from those encountered by the SSCs and, therefore, further tests on pilot plant scale are advised to simulate the SSCs to be decontaminated as closely as possible.

Important parameters to consider in pilot plant tests include the following: surface area to volume ratio (ratio of the surface area of the facility or items to be decontaminated to the volume of decontaminant); surface state of the materials (including any preconditioning before contamination); contact times of the various decontaminants employed; rinsing times; temperatures; turbulence (through stirring and agitation in laboratory tests; through proper flow rates and ad hoc arrangements of the samples in pilot plants); and dissolved oxygen.

6.5. COMMISSIONING

Commissioning ('cold commissioning') will be undertaken at several stages during the manufacture and installation prior to active operation of large decontamination facilities.

Commissioning typically includes the following stages:

(a) Inspection and testing at suppliers — undertaken by equipment suppliers of individual items of equipment.
(b) Trial assembly at works — undertaken by the manufacturer during the assembly of process plant modules containing several components, such as pumps, valves, instrumentation and pipework.
(c) Final acceptance testing at works — undertaken by the plant and equipment manufacturers to demonstrate a limited set of performance parameters of the part or wholly assembled plant modules. At this stage, some operator training may also be possible, along with testing and proving operating and maintenance instructions for plant items.
(d) Installation inspection and testing at site — undertaken to demonstrate correct construction of the plant and equipment, including non-destructive testing of welds, hydrostatic pressure testing and electrical testing (point to point, continuity).
(e) Site setting to work — once all relevant construction certification has been completed and inspected the site setting to work can be carried out, allowing the system to be cold commissioned and all functions of the entire system tested. This will include:
    (i) Connection to the control system.
    (ii) Vessel and pipework flushing to ensure they are clean.
    (iii) Pump running checks.
    (iv) Instrumentation checks.
    (v) Valve exercising.

(vi) Hard wired interlock testing.

(vii) Manual setting of valves followed by locking into position.

(f) Functional testing at site — the aim of the tests at site is to demonstrate the process operation of all items of plant and equipment working together following installation on the site (demonstration of the operation to be performed under the same operating regime (pressure, flow) as expected at the actual plant). These tests also test the control system. Site non-active functional testing involves a comprehensive test of modes of normal operation, operating and protective interlocks, fault recovery and maintenance operations.

(g) Inactive commissioning at site — this stage provides a demonstration of the whole plant. The final site inactive commissioning tests re-prove items tested at the off-site functional acceptance tests, although now under final plant conditions rather than with the temporary plant connections that were used for the work tests. The inactive commissioning may be carried out with water and, if required, an inactive simulant. Again, as for work testing, site inactive commissioning includes:

(i) Functional testing.

(ii) Performance testing.

Demonstration of the overall performance of the plant, with a high level of performance monitoring across the plant, together with health physics surveys, is undertaken to establish detailed operating parameters during inactive commissioning.

On completion of inactive ('cold') commissioning and the necessary safety documentation, the system is ready for active ('hot') commissioning or the operational phase.

6.6. OPERATION

Operational readiness actions (i.e. startup testing, design validation, plant familiarization by maintenance personnel) need to be completed prior to operational startup. Operational readiness not only requires operators to be trained in technology procedures, but also informational knowledge of the non-radiological environment, equipment and system interfaces.

6.6.1. Staffing

The types and extent of expertise required to deploy a decontamination facility vary widely with the technology used and the complexity of the process. In some instances, the appropriate staff are available at the site. For the implementation of a new process, the following considerations will be necessary:

— Training and qualification;
— Revisions of radiological and chemistry monitoring programmes;
— Revisions of equipment operation and maintenance;
— Procedural changes;
— Task scheduling.

After the appropriate training, it is prudent to perform a cursory evaluation to determine competence and the technical skill sets. This involves an assessment of the staffing and knowledge requirements associated with at least the following disciplines and activities:

— Equipment operation;
— Chemistry and radiochemistry sampling and analyses;
— Preventative, predictive and corrective maintenance;

— Radiological protection, including routine surveillance, personnel and area dose and dose rate monitoring, monitoring equipment maintenance and calibration, support of waste handling and packaging, system maintenance, and area and equipment decontamination activities;
— Secondary waste handling, treatment, conditioning and shipment or storage;
— Administration, including data analysis and facility, regulatory or other reporting;
— Management system quality inspections, testing and audit;
— Training type, frequency and content;
— Security and physical protection.

In addition to the above disciplines, the system process rate and frequency for related activities (sampling and analysis, preventative maintenance and testing) need to be defined so that the level of effort required to support those activities can be factored into the staffing analysis.

The costs for maintaining or leasing qualified personnel for operation and maintenance need to be considered. Further, in some instances, supplier based technicians for fixed or mobile equipment may be more experienced and better able to provide consistent, cost effective results than those achievable using facility staff who rotate through the operating positions and have collateral duties.

6.6.2. Procedures and programmes

After identifying tasks and staffing levels, procedures and management programmes are required to ensure that the overall processing programme is maintained at a high level of excellence. Existing facility procedures, equipment supplier manuals and industrial benchmarking could form the basis for procedures and management programmes. As a minimum, procedures for standard and non-standard (emergency actions) operations as well as maintenance and testing procedures need to be developed. For complex systems and/or mature programmes, additional procedures will be developed to address performance monitoring, reporting, management system review and oversight, and general organizational structures, roles and responsibilities.

6.6.3. Training

Regardless of the assumed familiarity with a technology, training needs to be conducted for essentially all systems prior to operation. The rapid rate of technological advancement makes it impractical for any individual to remain abreast of the required knowledge to operate new equipment safely and efficiently.

The training programme needs to be formalized to include, at a minimum, a training curriculum, training and qualification matrices, and archived documentation of the training given. Several industrial organizations (e.g. the World Association of Nuclear Operators (WANO) and the Institute of Nuclear Power Operations (INPO)) have developed very comprehensive guidance for training and qualification programmes, targeting sustainable operational excellence. The collective training basis and documentation is useful to determine who is qualified and competent to operate what, when they may need refresher training, and to provide proof of training in cases of litigation.

6.6.4. Process optimization

There are several avenues for optimizing a process. The most predominant methods include:

(a) Periodic assessments of the integrated management system — periodic evaluations that identify deficiencies and positive aspects of the process. These typically result in defining corrective actions using a formal programme.

(b) Periodic safety evaluations — these may be incorporated into the assessments of the integrated management system and typically result in actions using similar processes.

(c)   Staff feedback — many facilities have established programmes that actively solicit feedback from the staff. The feedback is incorporated into a corrective action programme for resolution.

(d)   Industrial operational experience — periodic searches of operational experience to identify opportunities for improvement. Those searches may identify actions that mitigate the potential for future equipment failure, and eliminate or minimize the potential for negative feedback from periodic assessments and evaluations.

(e)   Industrial benchmarking — active solicitation of opportunities for improvement and lessons learned via remote communications, forums, or facility visits. This has proven to be an invaluable method for process optimization.

(f)   Review of technical publications — the IAEA, as well as some industrial organizations (e.g. WANO, INPO), produce a significant number of technical reference publications that support process optimization and are based on actual experience and/or technology development.

(g)   Support of vendors and suppliers — direct interface with equipment providers and other subject matter experts to evaluate current performance and identify process enhancements.

6.6.5.   Performance monitoring

Performance monitoring is an integral element of the successful application of any technology or process discussed in this publication. The intention is to ensure on a predetermined basis that the process performance meets design criteria and define any trends to demonstrate that it is not degrading and is in a safe condition. The required performance parameters need to be defined during the design and commissioning phase and be based on objectives, goals, and equipment and media supplier suggestions. Typical data or parameters that would be monitored for a processing system include:

— Operating parameters such as pressure, flow, temperature;
— Chemistry parameters such as chemical impurities, pH, conductivity, oil and grease;
— Activity and dose rate trends;
— Failure types and their frequency, mode and causal factors;
— Cost per unit volume treated.

In some instances, performance monitoring may be an integral component of the management system's ongoing quality validation and/or safety case evaluations. Similarly, some of the performance data may support development of required periodic reporting to facility management, regulatory bodies or other agencies. Consideration needs to be given to determining some lead indicators, rather than just employing reactive monitoring, as these will allow safety and performance issues to be identified before they cause system problems.

6.6.6.   Documentation

Operating documentation includes the capture, collation and archiving of information related to all aspects of the process operation. This includes, but is not limited to, the staff selection process, training and qualification records, performance monitoring, inspections and tests. Similarly, maintenance records, equipment trends and corrective actions need to be documented in detail. The records are invaluable for future reference for performance enhancement, troubleshooting, modification activities and, less frequently, litigation.

The duration for which records need to be maintained varies by document type and the use of information as a reference and/or as an operational requirement. The term is typically defined in the facility licence and/or regulations. If no requirement exists, it would be prudent to maintain qualification records for staff for the duration of their employment, as well as records for performance, maintenance, inspection and testing throughout the entire decontamination process.

6.6.7. Risk management

Risk management of decontamination projects pursue goals of controlling and reducing the risk of injuries, errors, faults and accidents, so that workers and the public are not exposed to significant radiation, while at the same time improving quality. Risk management is a systematic process that involves the identification and assessment of risks followed by the elimination of risks in the first instance or, where this is not practicable, minimizing those risks as far as reasonably practicable. This enables the establishment of practice policies and procedures that can be implemented to control the risks that have been established as being 'likely' to happen. The risk assessment procedure can be summarized as indicated in Fig. 11.

A widely used technique to assess hazards is the so called hazard and operability study (HAZOP). HAZOP is a structured examination of a complex process or operation; it is intended to identify and assess issues that may incur risks to personnel or property. The objective of a HAZOP is to review the operation (here decontamination) to spot design and engineering issues that might otherwise have been lost. The technique is based on breaking the process into many simpler segments ('nodes'), which are then reviewed individually. A HAZOP study is performed by an experienced multidisciplinary team through dedicated meetings. HAZOP studies are qualitative and are intended to stimulate the imagination of the attending experts to discover hazards and operability issues. A more detailed overview of commonly used hazard evaluation techniques is available in Ref. [30].

HAZOP studies are best suited to assessing hazards in facilities, equipment and processes from multiple perspectives:

— Identifying weaknesses in design systems;
— Assessing an environment to ensure that a system is appropriately situated, supported, serviced and contained;
— Assessing engineered controls, sequences of operations and procedural controls (e.g. human interactions).

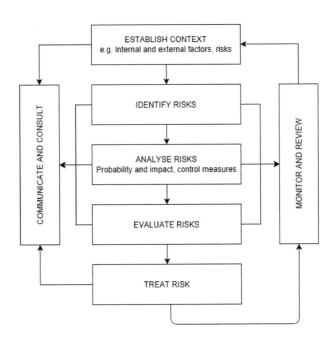

FIG. 11. Scheme of a risk assessment procedure.

6.7. DEMONSTRATION OF SUCCESS AND VALIDATION OF PROCESSES

Validation is a quality process establishing evidence that provides a high degree of assurance that a product, service, or system accomplishes its intended requirements. Post-decontamination monitoring for compliance with decontamination objectives is the final step in the process. This step is crucial if the objective is unrestricted release. At this stage, lack of compliance with decontamination criteria would mean repeating decontamination, perhaps by evaluating, planning for and applying an entirely different process. Significant delays, extra doses to the workers, double handling of the waste and extra costs could be expected.

6.8. REPORT TO REGULATORY BODIES AND OTHER STAKEHOLDERS

The final report of the decontamination event will include, among others, the type of radioactivity (alpha or beta–gamma) encountered, individual and collective doses, instruments used, time and location of the measurements, state of the examined material, what procedures have taken place, how long the decontamination procedure took and the total cost incurred.

The decontamination report is invaluable for the operating organization in that it proves successful and cost effective completion of the project, but it is also essential for the regulatory body as a demonstration that the project has been completed safely. The report can further be a valuable reference for planned or future decontamination projects and act as an aide-mémoire for the purposes of learning from experience.

7. DECONTAMINATION SELECTION ACTIVITY AS PART OF ACCIDENT MANAGEMENT

Routine decontamination activities could be defined as those activities occurring repeatedly during operation or maintenance. The scope of the decontamination work is well known and well understood. Non-routine decontamination activities could be defined as those occurring infrequently requiring a specific intervention, for example during a large accidental spillage or a laboratory spillage needing prompt decontamination.

7.1. DECONTAMINATION FOLLOWING LARGE SPILLAGE

Serious large volume spillages give rise to unanticipated and unplanned radioactive wastes, which require immediate management for protection of the public and the environment. In planning and considering measures for decontamination as a result of a nuclear spillage, the following parameters need to be assessed as far as practicable:

— Waste characteristics: physical properties, types of waste, levels of radioactivity, criticality risks and associated hazards (chemical, physical and other types);
— Waste location;
— Volume of radioactive wastes;
— Containment needed (avoiding the spread of radioactive substances during decontamination operations);
— Isolation required: reducing the radiation exposure from other radioactive waste;

— Distance and shielding: reducing the radiation exposure to decontamination workers;
— Time: reducing the radiation exposure time of workers.

Time is an essential parameter as radioactive waste arising from a large spillage is likely to include both fixed and non-fixed radioactive contamination from significant to very high levels of radiation. To prevent secondary exposure to the public and workers, it is necessary to implement emergency decontamination procedures promptly. The construction of a shielded area that provides shielding and measures to reduce the radiation exposure of workers includes setting up provisional shielding walls (concrete blocks used in generic civil engineering works), shortening the work time, clear delineation of the storage area and implementation of special procedures for waste handling, emplacement and monitoring.

In planning the decontamination of emergency wastes [31, 32], appropriate technologies need to be selected to reduce the radiation exposure of the public and workers on-site and off-site. However, the changing nature of radiological hazards over time may require the adoption of different decontamination methods and techniques.

The following steps can be followed as part of a decontamination selection activity:

(a) *Estimation/assessment of condition of radioactive wastes (including all SSCs)* — the location, volume, radioactivity and type of radioactive wastes within appropriate categories need to be estimated and assessed in the light of decontamination options.

(b) *Possibility of using readily available decontamination methodologies* — this needs to be assessed based on the estimated decontamination requirements for emergency wastes and the potential use of available decontamination techniques. Otherwise, decontamination options have to be explored elsewhere, which may take a long time.

(c) *Planning temporary storage facilities for decontamination waste* — if the existing radioactive waste storage facilities do not have enough capacity or no interim storage options exist, temporary storage facilities need to be planned and implemented based on the estimates and/or assessments made in step 1 and time constraints as in step 2.

(d) *Construction and operation of temporary storage facilities* — the storage facilities need to be constructed and operated according to the defined plans from step 3, based on the time requirements determined in step 2.

It is essential to systematically collect all lessons resulting from actual incidents and accidents, and circulate the information to the nuclear industry at large. For example, a database centred on tools (remotely controlled operations, robots) and recovery techniques (chemical decontaminants) would be invaluable.

7.2. DECONTAMINATION FOLLOWING SMALL SPILLAGE

Decontamination can be viewed as an impromptu, unplanned activity in the case of minor spillages in a laboratory or other small facility. In these cases, there is no need for the complex set of assessments, tests and decisions as simple decontamination methods are generically prescribed in laboratory manuals or left to skilled staff to determine. Belated intervention may incur penetration of contaminants through floor crevices under the floor coverings and make decontamination much more difficult.

8. CONCLUSIONS

There is no universal decontamination process that is applicable to all situations. Whilst there are a vast number of techniques available to choose from, selection of the approach best suited for a specific application requires plant and project specific assessment of technical, financial and commercial factors. Decontamination methods need to be selected based on effectiveness, compatibility with the system or component, secondary waste generation and waste disposal, as well as radiological and industrial safety. For the selection of a preferred decontamination process, the optimal balance needs to be established between all these factors. In some cases, decontamination may only be of marginal benefit and its costs may offset the benefits: if this is the case, the best approach may be not to decontaminate, but to dispose of directly.

Radiological and physical surveys of the SSCs to be decontaminated are crucial to appreciate and record radiation and/or contamination levels and physical conditions prior to commencing the operation. High radiation areas and areas where radioactivity is expected to build up during decontamination have to be identified. Such surveys not only familiarize the decontamination team with the plant layout and SSCs, but also contribute to the selection of decontamination approaches.

Training is instrumental to the success of a decontamination project. The training could prepare the decontamination team for the safe and cost effective implementation of the selected approach and the suitable use of PPE and contingency arrangements. Like the other aspects of decontamination planning and implementation, the objectives of the training programme need to reflect the plant specific radiological and industrial safety circumstances.

Often a combination of decontamination technologies is needed. Current decontamination techniques are case specific, and efforts could be made to broaden the scope of existing processes. To date, a significant amount of experience has been accumulated and the lessons learned implemented in the facilities. The main part of this publication presents many real case studies, which are further supplemented by examples in the annexes. While innovative technologies are constantly emerging, their limited technological readiness and the lack of large scale application are to be considered during the selection process. In all cases, health, safety and security are the main considerations when selecting, deploying and analysing the success of a decontamination process.

REFERENCES

[1] INTERNATIONAL ATOMIC ENERGY AGENCY, IAEA Nuclear Safety and Security Glossary, Non-serial Publications, IAEA, Vienna (2022).

[2] INTERNATIONAL ATOMIC ENERGY AGENCY, Policies and Strategies for Radioactive Waste Management, IAEA Nuclear Energy Series No. NW-G-1.1, IAEA, Vienna (2009).

[3] INTERNATIONAL ATOMIC ENERGY AGENCY, Leadership and Management for Safety, IAEA Safety Standards Series No. GSR Part 2, IAEA, Vienna (2016).

[4] INTERNATIONAL ATOMIC ENERGY AGENCY, Safety Assessment for Facilities and Activities, IAEA Safety Standards Series No. GSR Part 4 (Rev. 1), IAEA, Vienna (2016).

[5] INTERNATIONAL ATOMIC ENERGY AGENCY, Predisposal Management of Radioactive Waste, IAEA Safety Standards Series No. GSR Part 5, IAEA, Vienna (2009).

[6] INTERNATIONAL ATOMIC ENERGY AGENCY, Predisposal Management of Radioactive Waste from Nuclear Power Plants and Research Reactors, IAEA Safety Standards Series No. SSG-40, IAEA, Vienna (2016).

[7] INTERNATIONAL ATOMIC ENERGY AGENCY, Predisposal Management of Radioactive Waste from Nuclear Fuel Cycle Facilities, IAEA Safety Standards Series No. SSG-41, IAEA, Vienna (2016).

[8] INTERNATIONAL ATOMIC ENERGY AGENCY, Storage of Radioactive Waste, IAEA Safety Standards Series No. WS-G-6.1, IAEA, Vienna (2006).

[9]   INTERNATIONAL ATOMIC ENERGY AGENCY, Application of the Concepts of Exclusion, Exemption and Clearance, IAEA Safety Standards Series No. RS-G-1.7, IAEA, Vienna (2004).

[10]   INTERNATIONAL ATOMIC ENERGY AGENCY, Leadership, Management and Culture for Safety in Radioactive Waste Management, IAEA Safety Standards Series No. GSG-16, IAEA, Vienna (2022).

[11]   INTERNATIONAL ATOMIC ENERGY AGENCY, UNITED NATIONS ENVIRONMENT PROGRAMME, Regulatory Control of Radioactive Discharges to the Environment, IAEA Safety Standards Series No. GSG-9, IAEA, Vienna (2018).

[12]   INTERNATIONAL ATOMIC ENERGY AGENCY, Classification of Radioactive Waste, IAEA Safety Standards Series No. GSG-1, IAEA, Vienna (2009).

[13]   INTERNATIONAL ATOMIC ENERGY AGENCY, The Safety Case and Safety Assessment for the Predisposal Management of Radioactive Waste, IAEA Safety Standards Series No. GSG-3, IAEA, Vienna (2013).

[14]   INTERNATIONAL ATOMIC ENERGY AGENCY, Decommissioning of Nuclear Power Plants, Research Reactors and Other Nuclear Fuel Cycle Facilities, IAEA Safety Standards Series No. SSG-47, IAEA Vienna (2018).

[15]   INTERNATIONAL ATOMIC ENERGY AGENCY, Decommissioning of Facilities, IAEA Safety Standards Series No. GSR Part 6, IAEA, Vienna (2014).

[16]   INTERNATIONAL ATOMIC ENERGY AGENCY, State of the Art Technology for Decontamination and Dismantling of Nuclear Facilities, Technical Reports Series No. 395, IAEA, Vienna (1999).

[17]   INTERNATIONAL ATOMIC ENERGY AGENCY, Decommissioning of Research Reactors: Evolution, State of the Art, Open Issues, Technical Reports Series No. 446, IAEA, Vienna (2006).

[18]   INTERNATIONAL ATOMIC ENERGY AGENCY, Decontamination of Transport Casks and of Spent Fuel Storage Facilities (Proceedings of a Technical Committee Meeting, Vienna, 4–7 April 1989), IAEA-TECDOC-556, IAEA, Vienna (1990).

[19]   INTERNATIONAL ATOMIC ENERGY AGENCY, New Methods and Techniques for Decontamination in Maintenance or Decommissioning Operations, IAEA-TECDOC-1022, IAEA, Vienna (1998).

[20]   INTERNATIONAL ATOMIC ENERGY AGENCY, Spent Fuel Storage and Transport Cask Decontamination and Modification, IAEA-TECDOC-1081, IAEA, Vienna (1999).

[21]   INTERNATIONAL ATOMIC ENERGY AGENCY, Methods for the Minimization of Radioactive Waste from Decontamination and Decommissioning of Nuclear Facilities, Technical Reports Series No. 401, IAEA, Vienna (2001).

[22]   INTERNATIONAL ATOMIC ENERGY AGENCY, Decontamination Approaches during Outages in Nuclear Power Plants — Experiences and Lessons Learned, IAEA-TECDOC-1946, IAEA, Vienna (2021).

[23]   UNITED STATES ENVIRONMENTAL PROTECTION AGENCY, Technology Reference Guide for Radiologically Contaminated Surfaces, Report No. EPA-402-R-06-003, Washington, D.C. (2006).

[24]   OECD NUCLEAR ENERGY AGENCY, Decontamination Techniques Used in Decommissioning Activities: A Report by the NEA Task Group on Decontamination, OECD, Paris (1999).

[25]   INTERNATIONAL ATOMIC ENERGY AGENCY, Management of Low and Intermediate Level Radioactive Wastes with regard to their Chemical Toxicity, IAEA-TECDOC-1325, IAEA, Vienna (2002).

[26]   INTERNATIONAL ATOMIC ENERGY AGENCY, Fundamental Safety Principles, IAEA Safety Standards Series No. SF-1, IAEA, Vienna (2006).

[27]   OECD NUCLEAR ENERGY AGENCY, Remote Handling Techniques in Decommissioning, A Report of the NEA Co-operative Programme on Decommissioning (CPD) project, NEA/RWM/R(2011)2, OECD, Paris (2011).

[28]   Convention for the Protection of the Marine Environment of the North-East Atlantic (the OSPAR Convention), OSPAR Commission, London (1992).

[29]   EUROPEAN COMMISSION, Decision C(2014)4995, HORIZON 2020 — Work Programme 2014–2015.

[30]   UNITED KINGDOM HEALTH AND SAFETY LABORATORY, Review of Hazard Identification Techniques, Report No. HSL/2005/58, HSL, Sheffield (2005).

[31]   INTERNATIONAL ATOMIC ENERGY AGENCY, Management of Large Volumes of Waste Arising in a Nuclear or Radiological Emergency, IAEA-TECDOC-1826, IAEA, Vienna (2017).

[32]   INTERNATIONAL ATOMIC ENERGY AGENCY, The Fukushima Daiichi Accident, Report by the Director General, IAEA, Vienna (2015).

LIST OF ABBREVIATIONS

ALARA	as low as reasonably achievable
CORD	chemical oxidation reduction decontamination
CVCS	chemical volume control system
DF	decontamination factor
EIA	environmental impact assessment
EHS	environmental, health and safety
HAZOP	hazard and operability study
HEPA	high efficiency particulate air
ILW	intermediate level waste
LLW	low level waste
NPP	nuclear power plant
PPE	personal protective equipment
SSCs	structures, systems and components
TRA	technology readiness assessment
TRL	technology readiness level
UHP	ultra-high pressure
WAC	waste acceptance criteria

Annex I

LESSONS LEARNED FROM DECONTAMINATION

I–1. OVERVIEW

Contamination of materials and components occurs because of various physical and chemical processes. Contamination depth in non-porous metal is often much smaller than in building structures such as concrete, where it could be a few centimetres in thickness.

The amount of deposition of contaminants on the surface depends on several factors, such as the type of the base material, the surface roughness, the degree of corrosion and the material porosity, as well as the physicochemical properties of the fluid, such as pressure, temperature, pH value and contact time.

Depending on all these factors, the decontamination process can be different or can have varying results. Moreover, the objective of reduction in the level of contamination can also influence the selection of the process to be used. This objective can be one or more of the following:

(a) A reduction of the radiation dose rate of the component;
(b) The removal of loose or semi-loose contamination before dismantling or opening of a component to avoid a further risk of spread of contaminants during subsequent dismantling;
(c) The decategorization of the waste generated, or even the free release of the treated material, if the free release is an accepted and regulated process in the country where it is applied;
(d) The performance of a maintenance or decommissioning programme.

The potential or expected results and advantages of decontamination could be weighed against the internal and external dose accrued, the secondary waste generated and the cost of the operation itself. Further information can also be found in the decommissioning wiki of the IAEA's International Decommissioning Network programme at the IAEA website[1]. The following examples of lessons learned from decontamination activities include a brief technical description of the problems encountered. Specific lessons learned were identified by the authors or were easily derived from published material. Some of the material is anecdotal, but it was derived from those with direct experience.

This information does not necessarily reflect best practice and no judgement is made on the situations described. These are typical examples of the issues that can arise in the planning or implementation of decontamination activities. The information presented is not intended to be exhaustive and the applicability of these cases to a specific project needs to be considered. A general categorization of events is given in Table I–1.

I–2. CASE STUDIES

I–2.1. Unplanned exposure during fuel pool decontamination project, West Valley Nuclear Services, USA

I–2.1.1. Problem encountered

On 10 May 2002, at West Valley's New York facility, a lift rack was being removed from the fuel pool, when a layer of loose sediment was noted in the bottom channel of the lift rack [I–1]. An operator used the utility hose to spray out the sediment above the pool water surface. The spray was directed

---

[1] See: https://nucleus.iaea.org/sites/connect/IDNpublic/Pages/IDN-Wiki-Introduction.aspx

TABLE I–1. CASE STUDIES REPORTED IN THIS ANNEX

Lessons learned: categories	Reference in this annex
Unplanned exposure during decontamination	[I–2.1, I–2.3]
Spreading contamination	[I–2.4, I–2.13, I–2.17]
Secondary waste issues	[I–2.4]
Filter and ion exchange issues	[I–2.4, I–2.5]
Personal protective equipment	[I–2.6]
Industrial safety	[I–2.5, I–2.8]
Floor removal	[I–2.10]
Removal of radioactive oxides	[I–2.2, I–2.11]
Fissile material	[I–2.12]
Innovative technologies	[I–2.7, I–2.14, I–2.17, I–2.18, I–2.21, I–2.22]
Decontamination of hot spots	[I–2.2, I–2.15]
Alternative approaches to decontamination	[I–2.9, I–2.14, I–2.16]
Simple approaches	[I–2.19, I–2.20]
Backfitting to reduce source terms	[I–2.23]
Fogging as decontamination technology	[I–2.24]
Remote decontamination technology	[I–2.25]

towards the bridge crane operator who was lifting the rack. As was later noted, the recordings indicated an increase of air contamination in all three continuous air monitors around the time of the spraying. However, none of the continuous air monitor alarms were triggered, and the personnel contamination monitor detected no contamination. Two weeks later, during a routine total body count, the radiation protection technician who worked in the fuel pool was discovered to be internally contaminated with ^{137}Cs. Shortly after, an operator who had worked in the fuel pool during this period was also discovered to have inhaled ^{137}Cs. The data indicated that the two operators had received effective doses of 2 mSv and 1.65 mSv, respectively, from the uptake of airborne contamination (both values are still below West Valley's annual administrative control level of 5 mSv).

I–2.1.2. Analysis

Because of the unique features of the event, an independent assessment was carried out to establish direct, contributing and root causes. The review focused on the activities in the fuel pool facility around the time the event occurred. It was ascertained that the exposures were most likely to have occurred during the removal of the lift rack and/or during decontamination of the pool gate (the week before), which both involved similar activities. The exposures occurred when other activities were agitating the pool (through sprinkling) or as components were lifted from the pool and being sprayed to remove sediment. This conclusion was also supported by physical and radiological characterization of the pool water.

In planning for pool decontamination, no thought was given to the possibility of contamination from resuspension of sediments on components being lifted from the pool. Therefore, there was no requirement in the work procedure clearly specifying that the handling of equipment to be decontaminated or the removal of sediment only be performed underwater. Underwater work would have much reduced airborne contamination.

I–2.1.3. Lessons learned

— Strengthen hazard evaluation and monitoring when changes in work conditions are anticipated (changes in pool water turbidity; handling components with design features that tend to accumulate sediments);
— Only allow use of low pressure spray to rinse (not decontaminate) components;
— Perform decontamination underwater, where applicable;
— Specify the use of respiratory protection for activities where pool water is disturbed (spraying and equipment handling).

Another example is the cleaning and decontamination of workplaces containing beryllium in Canada [I–2].

I–2.2. Experience with decontamination gel at Lawrence Livermore National Laboratory, USA

I–2.2.1. Problem encountered

A previously Pu contaminated glovebox that was used to cold roll plutonium metal was converted into a temporary store for ^{238}Pu samples [I–3]. In 1996, a noticeable spill of ^{238}Pu occurred in the box, increasing the contamination significantly. The previous ^{238}Pu contamination of the glovebox made decontamination of the spill much more difficult. The objective of the decontamination was to reduce the activity of the glovebox to a level compatible with low level waste (LLW) disposal. A commercially available strippable coating was used to remove the ^{238}Pu spill, but it was insufficient to decontaminate the box to the required targets. A commercial gel (Decon Gel 1101) was available that, when cured, allows significant removal of contamination from metal surfaces in a strippable film that can be easily disposed of as solid waste. Decon Gel 1101 was applied to the Pu contaminated glovebox to determine its efficiency in removing contamination from several surfaces in a unique and highly contaminated environment.

After the gel had cured, the film barrier provided on average >91% shielding from the radiation over all surfaces measured. This is not surprising, given that most of the radiation was alpha. However, as the gel forms an impermeable film, extra protection from resuspension and extremity dose is provided.

I–2.2.2. Solution

After one application and removal of the Decon Gel 1101, the activity measured on the floor was reduced by 57% while that on the window was reduced by 37%. The window was subject to a second gel application, which achieved 99.5% removal of all activity. Similarly, the glovebox floor was subject to a second and third application and removal of Decon Gel 1101, resulting in an overall 99.4% removal of all activity. Given the high contamination of the surfaces, this decontamination efficiency was considered excellent.

I–2.2.3. Lessons learned

The traditional strippable coating used to remove small amounts of ^{238}Pu from metal surfaces was insufficient to decontaminate a plutonium spill in a glovebox to the required target. A commercially available gel was capable of penetrating into tight and hard to reach spots and removing more contamination at each pass; it can be applied easily and reasonably quickly. It appears to be versatile enough to be used either as a stabilizer, a decontaminant, or a pre-use protective coating, and addresses a wide range of contamination situations.

I–2.3. Internal contamination of several workers while cleaning up equipment, Vandellos-1 nuclear power plant (NPP), Spain

I–2.3.1. Problem encountered

On 28 February 1996, while some pieces of equipment were being decontaminated for removal, aerosols containing alpha particles, such as ^{241}Am, were generated [I–4]. Despite using masks equipped with filters for iodine and particles, 9 out of 13 workers at the working place were contaminated; 4 received doses up to 40% of the annual limit of intake and 5 received a lower level of contamination. The event was detected four days later, when air sampling filters where analysed.

I–2.3.2. Analysis

According to Section 5.1.4 of the INES User's Manual [I–5], when radiological control procedures and managerial controls are inadequate and workers receive unplanned radiation exposures (internal or external), such events can be rated level 1. Given that significant procedural failures led to this event, it was finally rated level 1.

I–2.3.3. Lessons learned

The possibility of airborne generation of radioactive particles had not been fully investigated and the protective measures taken were inadequate.

I–2.4. Primary loop decontamination at BR-3, Mol, Belgium

I–2.4.1. Problem encountered

In 1991, decontamination of the primary loop of the BR-3 prototype pressurized water reactor was carried out using the chemical oxidation reduction decontamination (CORD) process.

The full system decontamination reduced radiation levels in the primary system by a factor of 10. After decontamination, the general dose rate in the containment building was 0.08 mSv/h, a level that allowed personnel access [I–6]. The total dose received for the decontamination process was 0.16 person Sv. It is estimated that decontamination allowed a dose saving for dismantling the primary loop of some 4.25 person Sv.

More secondary wastes (mostly ion exchange resins) were generated than anticipated because of higher than expected quantities of crud.

I–2.4.2. Lessons learned

CORD is a well established process; only a few minor mishaps were noted. However, a primary system in sound condition is a prerequisite; and accurate planning and experienced workers contributed to its success.

Together with the primary loop, the reactor and its internals were decontaminated. This facilitated the dismantling and allowed the disposal of certain activated reactor parts as LLW, which would have been impossible without decontamination. One drawback was that during later work the visibility of the fuel storage pool was compromised by the resuspension of remaining attached contamination introduced by the process.

The design of future reactors needs to include features intended to streamline reactor decontamination to avoid later activities in a high radiation field such as installation of decontamination piping and sampling tubes.

I–2.5. Decontamination for decommissioning at Connecticut Yankee NPP, Haddam Neck, Connecticut, USA

I–2.5.1. Problem encountered

The Connecticut Yankee NPP's management chose to carry out full system decontamination prior to the dismantling of any primary components [I–7]. The goal was to decrease the dose rate by a factor of 15.

The decontamination strategy consisted of the removal of the reactor core barrel and the installation of a reactor nozzle dam system to reroute the flow from one steam generator loop to the other while bypassing the reactor pressure vessel. The circulation of the decontaminants was provided by the residual heat removal pumps operated in the shutdown cooling mode. The reactor coolant pumps were not used. This set-up allowed a high flow rate, which was sufficient to decontaminate all of the primary systems, such as the reactor coolant piping, residual heat removal system and chemical volume control system (CVCS), in one pass. The benefits of this strategy included:

— Reduced schedule because of a single pass for the entire system;
— Reduced waste volume;
— Removal of activated material from the reactor vessel.

Before the decontamination process started, laboratory tests were carried out on an artefact piping. The process optimization tests were targeted at comparing the effectiveness of various chemistries. These tests indicated that the Siemens HP CORD D UV[2] was the most effective method to remove the oxide films on Connecticut Yankee piping. The results of the tests showed that a significant amount of time and resources could be saved if chemical decontamination were carried out with some residual amount of boric acid. The flow test was conducted under similar conditions to those existing at Connecticut Yankee. During the tests, the amounts of dissolved activity and dissolved cations were measured in the decontamination solution to determine the decontamination factor.

The decontamination was planned and implemented making maximum use of the equipment installed at Connecticut Yankee, such as the residual heat removal pumps and the CVCS ion exchange columns. Prior to the actual decontamination, the systems were adapted for the decontamination process.

The primary system was drained and refilled prior to the first stage decontamination. Decontamination of the primary system was carried out excluding the reactor pressure vessel. To further reduce occupational doses, the decontamination solution was routed through the steam generator tubing instead of installing mechanical jumpers across the primary manways.

The decontamination process was initially scheduled to require four cycles of HP CORD D UV. However, owing to problems with the CVCS, the project was concluded after the second cycle. At that point in time, the average decontamination factor was 15.9. This was fully compatible with the selected dose reduction objectives.

---

[2] The acronym 'HP CORD D UV' stands for 'permanganic acid (HP), chemical oxidation reduction decontamination (CORD), decommissioning (D) and ultraviolet light (UV)'.

I–2.5.2. Lessons learned

During the decontamination process, it became evident that the CVCS ion exchange columns were unsuitable for decontamination application because of their age and physical condition. The problems included difficult valve operations and unexpected leaks. During the decontamination cycles, it was not possible to operate the CVCS ion exchange beds for the anticipated duration. Ultimately this led to disruption of solvent regeneration and activity removal, but it did not affect the overall solvent effectiveness, it merely caused a delay in the project schedule. This case proves that laboratory tests under standard conditions do not necessarily reproduce the real working environment.

I–2.6. Respirator reduction aids decontamination programmes

I–2.6.1. Problem encountered

A general problem in decontamination activities is the production of dust, leading to the spread of airborne contamination, with the resulting need for respiratory control devices. Concerns about excessive respirator use refer to workers breathing more deeply to obtain oxygen, which contributes to dizziness, accelerated fatigue and heat stress. The US Department of Energy's (DOE's) Savannah River Site has implemented a respirator use reduction programme, and has sought out more effective engineering controls to reduce reliance on respirators. A contractor was selected to provide effective decontamination services whilst minimizing the generation of airborne contamination, effectively reducing the need for respiratory protection.

The contractor's dustless decontamination system removes the health concerns of respirator overuse and permits longer working times. The success of the respirator use reduction programme relies heavily on the engineering controls used to reduce airborne contamination.

The Indian Point NPP deployed another dustless decontamination technique. The contractor's decontamination team was tasked to remove paint and concrete substrate in 16 areas totalling ~1600 m^2 with the aim of releasing these areas from regulatory control.

Most of the flooring was flat and unobstructed — ideal for the contractor's MOOSE, a remotely operated floor scabbling robot with an onboard vacuum cleaner equipped with a high efficiency particulate air (HEPA) filter. MOOSE collects and packages waste in a single step dustless procedure.

I–2.6.2. Lessons learned

As respirators can increase external exposure by affecting worker efficiency, their use needs to be minimized [I-8]. Concerning the removal of hot particles from nuclear primary heat transfer systems using custom designed robots, see also Ref. [I–9].

I–2.7. Experience of laser decontamination in Brazil

I–2.7.1. Problem encountered

Between 1970 and 1980, radioactive lightning rods containing ^{241}Am sources were used in Brazil. Eventually these devices were found to be generally ineffective and the Brazilian regulatory body suspended their use and they were shipped to a centralized radioactive waste management facility for processing. The first treatment step was the removal of the radioactive sources. However, some americium contamination remained and decontamination was needed to prevent the material from being dealt with as radioactive waste [I–10].

I–2.7.2. Analysis

There are several methods for removing contaminated metal oxide layers, but in this case the Brazilian authorities decided to use laser decontamination methods. It allowed the recycling of the decontaminated radioactive lightning rods, reducing radioactive waste volumes and management costs.

I–2.7.3. Lessons learned

In this case, although there were many well established decontamination techniques, research and development activities are producing many new methods. The development of such new techniques may have significant advantages over other more traditional ones. Laser decontamination is one of the emerging techniques to have gained a following in recent years. The advantages of laser decontamination include:

— The low impact on work environment;
— The small quantity of secondary waste;
— Remote applicability (with a resulting reduction of occupational exposures);
— Decontamination equipment not exposed to contamination;
— The possibility of automation.

One disadvantage is the required accuracy of the laser beam focusing on the surface being treated and the requirement for a three phase power supply. The operation of the laser itself also requires a strict maintenance schedule, as the interaction of air on the optic surface causes deterioration

I–2.8.   Arc flash accident at Los Alamos National Laboratories, New Mexico, USA

I–2.8.1. Problem encountered

During preventive maintenance on an electrical substation, an employee entered a cubicle on the energized portion of the switchgear to clean it with a commercial spray cleaner. The worker suffered severe injuries from the resulting arc flash and blast. The cleaning solution created a discharge path between the 13.8 kV transformer and the grounded cubicle. The force of the arc flash and blast ejected the worker from the cubicle. In addition to suffering significant burns, the worker lacerated his head when he fell backward and struck nearby test equipment [I–11].

It was disclosed that the spray cleaner was intended for use only on non-energized surfaces because it has no established insulating properties to prevent conduction of an electrical current.

I–2.8.2. Lessons learned

The review team identified many shortcomings that are broader in scope than this electrical incident. However, the most significant one from the decontamination point of view is that management needs to reinforce and clarify expectations and implementation of zero voltage verification requirements during maintenance (decontamination).

I–2.9.   Cost–benefit evaluation of decontamination a facility at Windscale Advanced Gas Cooled Reactor, United Kingdom

I–2.9.1. Problem encountered

When decommissioning the Windscale Advanced Gas Cooled Reactor, various tests were undertaken to assess the applicability of the heat exchanges to decontamination processes. Nitric acid (0.5M) with a small addition of citric acid (0.0025M) was found to be most effective, as it would have the least impact on the effluent treatment system [I–12].

I–2.9.2. Analysis

Trials involving the spraying of the acid into one section of the heat exchanger (applied to avoid generating large volumes of secondary waste) achieved a decontamination factor of ~3. In some circumstances this might be acceptable, but in this case it was judged to be too small. Hence disposal without decontamination was adopted and the heat exchangers were removed and shipped intact to the LLW repository.

I–2.9.3. Lessons learned

The solution identified in trials as the most applicable decontamination method is not always sufficient in achieving the desired outcome. In such cases, the cost of undertaking a decontamination trial by far outweighs any benefits realized and the most appropriate course of action would be to dispose of the items without decontamination.

I–2.10. Rehabilitation of a low level radiologically contaminated flooring system

I–2.10.1. Problem encountered

Facilities at the NPP were used to store and decontaminate equipment. Over time, some radiological contamination had attached to the floor surfaces. To prevent this contamination from becoming airborne, a remediation strategy was planned to safely remove the existing floor and replace it with a high density impermeable surface

I–2.10.2. Analysis

The strategy included the removal of the uppermost 12 mm of the floor. The floor was ribbon cut with a customized diamond gang saw and all materials removed (concrete and slurry) were contained and packaged for final disposal. The removal operations were carried out in a specially designed controlled negative air environment, ventilated through HEPA filters. The workers wore air supplied suits. Following decontamination, the floor was cleaned mechanically using self-propelled shot blasting equipment. A high density epoxy mortar overlay with fiberglass reinforcing was placed over the entire area. The new floor provides extremely low permeability.

I–2.10.3. Lessons learned

Preventive action needs to be undertaken to reduce the likelihood of contamination becoming entrained within flooring material. This includes the use of impermeable floor surfaces and the routine removal of any loose contamination. This prevents the need to routinely replace old floors and reduces the amount of secondary waste generated.

I–2.11. Removal of radioactive oxide buildup from reactor coolant pipelines

I–2.11.1. Problem encountered

Nearly 9 m^2 of highly radioactive oxide buildup (80–120 mSv/h) on stainless steel reactor coolant piping in a US NPP needed to be removed. Utility representatives requested that no appreciable loss of the stainless steel substrate occur and specified an overall surface profile of <60 μm. A contractor was tasked to remove the oxide layer with minimal substrate damage and safely decontaminate the pipe to below 100 000 dis/min (1660 Bq), so that other outage maintenance could be conducted [I–13].

I–2.11.2. Analysis

A dry abrasive technique was selected in this circumstance. The reasons for selecting this method included the need for low dust generation and a clean, dry process that generates low volumes of waste. Specifically, sponge blasting media were selected, which were propelled through the pipe.

I–2.11.3. Lessons learned

A dry abrasive technique was able to remove the oxide layer, reducing contamination levels and keeping the substrate profile as required. Other case studies are given in Refs [I–14 to I–16].

I–2.12. Robust estimation of fissile material quantity in a dissolver

I–2.12.1. Problem encountered

A continuous process was used for the dissolution of spent reactor fuel in a reprocessing plant. The basic decontamination parameters stemmed from values obtained many years before, when a dissolver was replaced and dose rate measurements were taken. A first rinse using nitric acid was conducted without any prior chemical analysis of the deposits at the bottom of the dissolver; the quantity of plutonium removed turned out to be much higher than the estimated amount. The use of incorrect initial parameters also affected the level of trust of the regulatory body supervising this project [I–16] and there was a considerable impact from the rescheduling of activities.

I–2.12.2. Analysis

The discrepancy in the estimation of fissile material in the dissolver, once it was discovered, led the operator to quickly redefine a characterization programme to determine the physical, chemical and radiological features of the deposit and the remaining quantity of plutonium. Non-destructive assays, such as active and passive neutron measurements, were used, as well as video and sampling analysis.

I–2.12.3. Lessons learned

The problem encountered during the rinsing of the continuous dissolver was a consequence of poor knowledge of the initial state, mainly resulting from a lack of characterization of the deposits inside the dissolvers. The approach taken was then to monitor the initial state of the dissolver and to characterize the deposits carefully. If done initially, this characterization could have allowed an accurate estimation of the amount of residual fissile material and adjustment of the decontamination solutions and set-up in response to the circumstances encountered.

I–2.13. Contamination caused by a leaking radioactive source, Cuba

I–2.13.1. Problem encountered

A small research facility was used to store spent sealed radioactive sources, in which a ^{137}Cs source leaked. Significant contamination was detected on some walls and floors in 1980. Owing to a lack of waste management expertise, infrastructure and financial resources, the contaminated areas were simply locked up and abandoned. In 1986, in a decontamination attempt, the walls and floor were washed using pressurized water jets. This method was ineffective, as the contamination was only reduced by ~20% and the contaminated water spread the contamination to clean areas [I–17].

I–2.13.2. Analysis

First, all stored radioactive waste was removed from the facility. Chemical decontamination of all surfaces was carried out. The waste solutions were collected and subsequently processed. Vacuum devices and polyurethane sponges were also used for recovering liquid waste. Part of the floor surface was removed from some rooms.

I–2.13.3. Lessons learned

In the first instance, the use of water jetting spread contamination further and was ineffective in achieving decontamination. In subsequent decontamination attempts, hydrochloric acid solutions containing potassium alum $(KAl(SO_4)_2)$ were used for clay or cement floors and walls. For soils, concrete and asphalt, Prussian blue $(Fe_4[Fe(CN)_6]_3)$ was added to the solution.

I–2.14. Metal decontamination, Hinkley Point A, United Kingdom

I–2.14.1. Problem encountered

Ultra-high pressure (UHP) water blasting was deployed at Hinkley Point A for the decontamination of a few pond skips. For some of the skips for which activity levels were low, UHP blasting removed sufficient amounts of contamination. The decontaminated skips were subsequently melted at the Energy Solutions Bear Creek Processing Facility, located near Oak Ridge, Tennessee, USA, enabling the beneficial reuse of metal. Other skips of higher activity were not in compliance with the melting facility's conditions for acceptance and, hence, an alternative decontamination technique was sought.

I–2.14.2. Analysis

Decontamination trials were conducted on a representative number of Magnox skips using UHP blasting and UHP blasting with abrasives. The trials demonstrated that the decontamination factors delivered by UHP blasting (and other techniques) were not high enough to reduce intermediate level waste (ILW) skips to LLW skips or LLW skips to metal melt acceptance criteria.

Based on the trial results, there was a significant risk that many ILW skips would remain ILW even after decontamination with UHP blasting, along with the risk of creating secondary wet ILW. Furthermore, a significant amount of worker dose would be expended to no avail. The initial trial also proved that a significant amount of the radioactivity was embedded in the base metal or driven into it by the decontamination processes.

The most promising technology that emerged from the trials was the milling of metal with standard industrial computer numeric controlled milling machines. The milling technology appeared to consistently achieve decontamination factors of 10 or greater on ILW skips, without generating secondary wastes additional to the actual material being removed.

I–2.14.3. Lessons learned

The ILW skip problem has shown that the constant improvement of technologies being used in non-nuclear industries can provide off the shelf products for solving problems within the nuclear decommissioning industry. The approach of decommissioning by trial has also proved to be an efficient way to minimize the safety and economic hazards inherent to new technologies by breaking down a difficult problem into smaller problems that can be solved through inexpensive small scale trials.

I–2.15. Unexpected need for decontamination, Trino NPP, Italy

I–2.15.1. Problem encountered

The removal of non-irradiated fuel elements, which had been placed in the reactor core but never exposed to the neutron flux, presented the opportunity to perform several checks on the elements when they were removed. The fuel elements were decontaminated, dried and transferred to the fresh fuel storage facility in a nylon container. During removal, surveys were conducted to locate any hot spots due to crud deposits on the element. This highlighted, in certain cases, a dose rate that was more than twice the one anticipated. [I–18].

I–2.15.2. Analysis

The higher dose rates were due to metal shavings on the elements, originating from 1968 when the thermal shield was cut. A purpose built nozzle with appropriate adaptors was manufactured. This made it possible to direct the water jet onto the shavings and remove them by flushing.

I–2.15.3. Lessons learned

Even the simplest operations can lead to unforeseen problems, originating from previous and almost forgotten operations. Assumptions regarding whether or not there is a need for decontamination have to be proven in all instances.

I–2.16. Cost–benefit analysis underpins the decontamination strategy, Miamisburg Environmental Management Project, Ohio, USA

I–2.16.1. Problem encountered

During decommissioning of Building 21 at the Miamisburg Environmental Management Project (MEMP), consideration was given to the benefits of undertaking a decontamination procedure [I–19].

I–2.16.2. Analysis

Even without a detailed cost estimate, existing information can be used during decommissioning projects to estimate the relative costs of each option through an order of magnitude cost–benefit analysis. With this method, decommissioning options that are not cost effective can be discarded from further consideration at an early stage of the planning. For Building 21 at the MEMP, a cost–benefit evaluation indicated that disposal of construction rubble as LLW with no decontamination was more cost effective than decontamination of the building followed by disposal of the debris as inactive waste. MEMP estimated that this approach could result in significant cost savings.

I–2.16.3. Lessons learned

Decontamination is not always mandatory. Sometimes there are factors that render decontamination impractical, and a robust cost–benefit analysis has to be undertaken before embarking on any remediation action.

I–2.17. Experience with vacuumable decontamination gel for surface decontamination at Marcoule, CEA, France

I–2.17.1. Problem encountered

Some hot cells inside the Facility for Post-irradiated Fuel Examination (ISAI) at the Marcoule Nuclear Research Centre were contaminated and required decontamination [I–20, I–21].

I–2.17.2. Analysis

Vacuumable decontamination gels are already used on a regular basis in nuclear decommissioning. A typical example of such gels is ASPIGEL 100E. This gel is based on an acidic Ce^{4+} containing formulation and is well suited for the removal of fixed contamination present in several tens of micrometres of the material. Another example is ASPIGEL 400, which is a basic formulation for aluminium alloy material decontamination.

The gel was sprayed to form a homogeneous layer of the desired thickness (0.5–1 mm). Upon drying (2–48 h, depending on the formulation and the climatic conditions), the gel shrunk and resulted in the formation of cracks and flaking. The flakes were easily removed using brushing and/or vacuuming.

Based on experience in the formulation of nuclear decontamination gels, the use of inorganic gels to decontaminate building materials or civil infrastructures in the frame of chemical, biological, radiological or nuclear (CBRN) events was investigated. The aim was to develop an easily deployable multitask formulation, which would be sprayable and vacuumable and would generate small volumes of easy to handle waste. This research was a collaboration between the French Atomic Energy and Alternative Energies Commission (CEA) and the company NBC-Sys (the corresponding project is named GIFT-RBC). Using the synergies between the two organizations, different vacuumable gels were developed to handle decontamination relating to CBRN events on various materials using adequate spraying and suction techniques.

I–2.17.3. Lessons learned

The gel is applied onto the contaminated surface. Using suited equipment, the operator can cover up to 4 m^2 of surface/min. During the drying process, the presence of the operator is not required in the room. After drying, the dry gel entraps the contaminants and can be removed by vacuum cleaning. The whole process considerably limits personal exposure and the secondary waste generated is in a solid form and is therefore easy to handle and manage for disposal. The gel formulation can be adapted for stainless steel or concrete application. This annex highlights the need to consider several factors in the selection of a decontamination process, such as the application method, the operator interfaces, the substrate material to be decontaminated and waste handling.

I–2.18. High alpha surface contamination, Mound facility stack, USA

I–2.18.1. Problem encountered

Areas of high alpha contamination entail occupational hazards and can render dismantling complicated [I–22].

I–2.18.2. Analysis

The contractor developed a fogging technology to apply an aerosol coating all over the contaminated surface of a stack. The application was possible without intruding into the contaminated area. Loose surface contamination was lowered by a factor of 1000 and more. The technique required minimal PPE and allowed dismantling to occur in a timely manner.

I–2.18.3. Lessons learned

As in previous cases, new technologies need to be considered while a project is still at a design stage.

I–2.19. Use of off the shelf products, Dounreay, United Kingdom

I–2.19.1. Problem encountered

Common chemical decontamination fluids are often ineffective in removing hot spots of surface contamination. These chemicals need time to dry and draw contamination out of any pores. However, the acids that had been used as decontamination agents in the past at Dounreay also created problems [I–23]. Therefore, it was necessary to identify a product that enables the effective and efficient removal of plutonium from steel before it can be disposed of.

I–2.19.2. Analysis

Dounreay workers discovered that an inexpensive household cleaner removes plutonium stains more effectively than many nuclear decontamination products. Use of this commercially available product (Cillit Bang) was approved by management and it was employed to remove small hot spots of activity. Subsequent tests confirmed its effectiveness.

I–2.19.3. Lessons learned

This case presents a good example of innovation pushing down the cost of decommissioning a nuclear site; the generation of secondary waste was minimal and the product was compatible with pre-existing waste acceptance conditions. The product could be deployed quickly and effectively as management readily accepted its use.

I–2.20. Simple decontamination techniques at BR-3, Mol, Belgium

I–2.20.1. Problem encountered

During the decommissioning of metallic components and concrete at the BR-3 prototype pressurized water reactor it was realized that it was necessary to remove contamination from many surfaces [I–24].

I–2.20.2. Analysis

One of the simplest and most inexpensive ways to decontaminate metallic components that have embedded contamination is to grind the surface using an electric grinder. The grinders used at BR-3 were off the shelf products designed for polishing. For decontamination of concrete, international best practice is often stated to be the use of diamond tipped tools, but simple techniques such as scabbling can also give acceptable results and the investment cost is much lower than for diamond tipped tools. Similarly, a simple air powered pneumatic drill can give good results in removing embedded components inside concrete blocks and in removing deep contamination.

I–2.20.3. Lessons learned

The BR-3 decommissioning project has proven that a large percentage of decontamination tasks can be safely carried out using conventional tools available in the non-nuclear market. Some of these tools require adaptation to be deployed remotely or to prevent the spread of contamination. Even where the costs of human resources are high, manual operations can often be more cost effective than automated ones. This is a fortiori more evident where the costs of human resources are low.

I–2.21. Decontamination process using aqueous static foams

I–2.21.1. Problem encountered

During chemical decontamination using a water jet, a significant volume of secondary waste is often generated, which requires treatment. Foam decontamination is advantageous for use in areas of complex geometry, pipework and small internal structures. Foam decontamination processes have been assessed as an alternative to liquid methods and have been deployed at the Phebus Fission Product Experimental Target Chamber [I–25, I–26] located at CEA's Cadarache Research Centre in France. Foam decontamination has also been used in projects at the Russian Federation's A.A. Bochvar High Technology Scientific Research Institute for Inorganic Materials (VNIINM).

I–2.21.2. Analysis

At VNIINM, the internal cavity of the object to be decontaminated was filled with foam and after introducing a foam breaking process the minimal generated solution containing the activity was removed. A range of specific foam forming solutions have been designed and tested for the removal of various types of radioactive contaminants. A mobile foam decontamination facility has also been developed. After several years of using the foam decontamination process, complex fluids have been developed (micellar solutions, supercritical fluids, gels and foams) that do not damage the surface of the materials (soft homogeneous corrosion) and rapidly transfer the radioactive contaminants into the fluid phase. These goals were achieved by adding small quantities of surfactants or polymers in the complex fluid phase.

I–2.21.3. Lessons learned

Decontaminating closed internal structures is often cumbersome and in many cases users have simply disregarded equipment and opted for disposal. Foam decontamination, as demonstrated in the example above, allows the user to consider decontaminating using foam instead of traditional water based applications.

I–2.22. Re-evaluation of a decontamination facility based on cost

I–2.22.1. Problem encountered

The Dresden 1 NPP is one example where an entire new system was constructed and installed for the decontamination of the primary system. Dresden Unit 1 had a history of minor steam leaks and erosion in steam piping. There were also fuel failures, which caused the redistribution of radionuclides from the fuel to other parts of the primary system. The use of Cu–Ni surfaces led to translocation and deposition of corrosion products throughout the operating systems [I–27, I–28]. The use of carbon steel in the secondary feedwater system may have also contributed to elevated corrosion radionuclide levels. These issues gave rise to the need for a planned chemical decontamination of the primary system.

I–2.22.2. Analysis

Dresden Unit 1 was taken off-line to backfit it with equipment to meet new federal regulations and to perform chemical decontamination of major piping systems. While it was out of service for retrofitting, additional regulations were issued because of the accident at Three Mile Island Unit 2 in March 1979. The estimated cost to bring Dresden Unit 1 into compliance with these regulations was excessive, and it was concluded that the age of the unit and its relatively small size did not warrant the added investment. In 1984, chemical decontamination of the primary system was performed and 28 TBq of ^{60}Co and 0.05 TBq of ^{137}Cs were removed.

I–2.22.3. Lessons learned

The need to treat large amounts of chemical wastes resulting from a large scale project requires the construction of an entirely new facility. Cost considerations may render the entire project unfeasible, although decontamination of small SSCs within the larger plant may still be viable.

I–2.23. Backfitting or decommissioning to reduce source terms

I–2.23.1. Problem encountered

Two early projects, Hot Spot Removal and Reactor Coolant System Decontamination, were initiated at the Maine Yankee NPP in preparation for decommissioning during the operation to decommissioning transition period [I–29]. Radiation surveys conducted during plant operation had noted a few hot spots located at typical activity buildup points such as piping elbows, valve connections and locations in piping with flow changes. The hot spot removal programme was meant to geographically identify the hot spots in detail to allow their 'surgical' removal (i.e. only cutting out the highly contaminated valve or piping section).

I–2.23.2. Analysis

The Maine Yankee NPP used a computerized gamma video camera to identify the contamination in situ while continuing operation. The visual image of the monitored zone in black and white had superimposed colour variations correlated to radiation exposure variations. The images allowed identification of the hot points, which were later removed. It was estimated that the hot spot removal programme reduced the decommissioning project exposure by ~150 person rem (1.5 person Sv).

In preparation for decommissioning, the Maine Yankee project also performed chemical decontamination of the reactor coolant system. It was estimated that this decontamination reduced the total project exposure by another ~150 person rem (1.5 person Sv).

I–2.23.3. Lessons learned

The task of reducing radiation source terms in preparation for activities such as backfitting or decommissioning can be pursued for a different objective than reducing the source terms. The objective could be to create a less contaminated environment in preparation for large scale human access; in the case of a pond or tank, minor amounts of sludge and contaminated water may remain.

I–2.24. Fogging as decontamination technology

I–2.24.1. Problem encountered

At the Humboldt Bay NPP, Eureka, California, USA, the main plant stack, a 90 m reinforced concrete structure, had to be dismantled due to seismic concerns. In addition to the stack, large sections of the ventilation system had to be removed. A contract was stipulated to encapsulate contamination within the ventilation system and the stack before dismantling.

I–2.24.2. Analysis

Contamination was fixed by fogging. The process uses a device producing an aerosol of capture coating. The aerosol condenses on surfaces, sequestering the contaminants in situ. The contractor used air filled balloons to seal off sections of the ventilation system while other parts of the system were being dismantled.

I–2.24.3. Lessons learned

Fogging is a patented process for eliminating airborne radioactivity and fixing contamination in situ remotely without the need for entering the contaminated area [I–30]. Use of this fogging technology also reduces or pre-empts the need for glove bags and extensive contamination control during dismantling and removal of contaminated components.

I–2.25. Remote decontamination technology

I–2.25.1. Problem encountered

The decommissioning of the first generation Magnox storage pond at Sellafield, United Kingdom [I–31], was problematic as the contaminated pond had a considerable associated radiation field, caused by the buildup of legacy radioactive material over decades. This radiation field prevented access of personnel to the close vicinity of the wall and had been the major obstacle in implementing a technical solution for the required remediation work.

I–2.25.2. Analysis

The solution for the problem came in the form of a powered remote manipulator arm, deployed by a mobile crane. In addition, various tooling and specialist resin technology provided the necessary capabilities to stabilize, isolate and remove the corroded pipework, as well as to decontaminate and seal the pond wall. The project, including decontamination aspects (the hydroblasting tool and effluent system), is described in detail in Ref. [I–31].

I–2.25.3. Lessons learned

The decommissioning project has proven that a large percentage of decontamination tasks can be safely carried out using remote techniques with low impact on exposure of workers.

REFERENCES TO ANNEX I

[I–1]   INTERNATIONAL ATOMIC ENERGY AGENCY, Decommissioning of Pools in Nuclear Facilities, IAEA Nuclear Energy Series No. NW-T-2.6, IAEA, Vienna (2015).

[I–2]   INSTITUT DE RECHERCHE ROBERT-SAUVÉ EN SANTÉ ET EN SÉCURITÉ DU TRAVAIL, Cleaning and Decontamination of Workplaces Containing Beryllium, Report R-614, IRSST, Montreal (2009), http://www.irsst.qc.ca/media/documents/pubirsst/R-614.pdf

[I–3]   SUTTON, M., FISCHER, R.P., THOET, M.M., O'NEILL, M., EDGINGTON, G., Plutonium Decontamination Using CBI Decon Gel 1101 in Highly Contaminated and Unique Areas at LLNL, Report No. LLNL-TR-404723, LLNL, Livermore, CA (2008).

[I–4]   LANDELIJK KERNENERGIE ARCHIEF, Internal Contamination of Several Workers While Cleaning up Equipment, Vandellos-1 NPP, Spain, LAKA, Amsterdam (1996), https://www.laka.org/docu/ines/event/562

[I–5]   INTERNATIONAL ATOMIC ENERGY AGENCY, INES: The International Nuclear and Radiological Event Scale User's Manual, Non-serial Publications, IAEA, Vienna (2013).

[I–6]   INTERNATIONAL ATOMIC ENERGY AGENCY, State of the Art Technology for Decontamination and Dismantling of Nuclear Facilities, Technical Reports Series No. 395, IAEA, Vienna (1999).

[I–7]   NUCLEAR ENGINEERING INTERNATIONAL, Decontamination for Decommissioning at Connecticut Yankee, NEI, London (1998), http://www.neimagazine.com/features/featuredecontamination-for-decommissioning-at-connecticut-yankee/

[I–8]   JOHNSON, A.T., Respirator masks protects health but impact performance: a review, J Biol Eng., **10** 4 (2016).

[I–9]   KINECTRICS, Hot Particle Removal Project (2016), https://www.kinectrics.com/capabilities/services/nuclear-equipment-development-supply/first-of-a-kind-robotics-development-testing

[I–10]   POTIENS, A.J., et al., Laser decontamination of the radioactive lightning rods, Radiat. Phys. Chem. **95** (2014) 188–190.

[I–11]   UNITED STATES DEPARTMENT OF ENERGY, OFFICE OF ENVIRONMENT, HEALTH, SAFETY AND SECURITY, Operating Experience Summary, Report 2016-01, USDOE, Washington, D.C. (2016), https://www.energy.gov/sites/prod/files/2016/01/f28/OES_2016-01.pdf

[I–12]   BAYLISS, C., LANGLEY, K., Nuclear Decommissioning, Waste Management, and Environmental Site Remediation, Butterworth-Heinemann (2003).

[I–13]   SPONGE-JET, INC., Removal of Radioactive Oxide Build-up from Reactor Coolant Piping, Sponge-Jet, Newington, NH (2013), https://www.spongejet.com/wp-content/uploads/2013/03/Removing_Radioactive_Oxide_Build-up_eng.pdf

[I–14]   SCHAUDER, B., Reactor head restoration, FirstEnergy Nuclear Operating Company, Sponge-Jet, Newington, NH (2013), https://www.spongejet.com/wp-content/uploads/2013/03/Davis_Besse_Reactor_Head_eng.pdf

[I–15]   SPONGE-JET, INC., Sponge Jet Decontamination System, Sponge-Jet, Newington, NH (2013), https://www.spongejet.com/wp-content/uploads/2013/03/DOE_Decontamination_Cost_Analysis_eng.pdf

[I–16]   INTERNATIONAL ATOMIC ENERGY AGENCY, Planning, Managing and Organizing the Decommissioning of Nuclear Facilities: Lessons Learned, IAEA-TECDOC-1394, IAEA, Vienna (2004).

[I–17]   INTERNATIONAL ATOMIC ENERGY AGENCY, Decommissioning of Small Medical, Industrial and Research Facilities, Technical Reports Series No. 414, IAEA, Vienna (2003).

[I–18]   INTERNATIONAL ATOMIC ENERGY AGENCY, Transition from Operation to Decommissioning of Nuclear Installations, Technical Reports Series No. 420, IAEA, Vienna (2004).

[I–19]   UNITED STATES DEPARTMENT OF ENERGY, OFFICE OF ENVIRONMENTAL MANAGEMENT, Decommissioning Handbook: Procedures and Practices for Decommissioning, Report No. DOE/EM-0383, USDOE, Washington, D.C. (2000), https://www.osti.gov/servlets/purl/1491121

[I–20]   FAURE, S., FUENTES, P., LALLOT, Y., Vacuumable Gel for Decontaminating Surfaces and Use Thereof, World Patent WO 07/039598 (2007).

[I–21]   CUER, F., FAURE, S., Biological Decontamination Gel, and Method for Decontaminating Surfaces Using Fluid Gel, World Patent WO 12/001046 (2012).

[I–22]   INTERNATIONAL ATOMIC ENERGY AGENCY, Dismantling of Contaminated Stack at Nuclear Facilities, Technical Reports Series No. 440, IAEA, Vienna (2005).

[I–23]   NUCLEAR DECOMMISSIONING AUTHORITY, "Cillit Bang will do the trick", Insight into Nuclear Decommissioning, NDA, London (2009).

[I–24]   INTERNATIONAL ATOMIC ENERGY AGENCY, Decommissioning Techniques for Research Reactors, IAEA-TECDOC-1273, IAEA, Vienna (2002).

[I–25] FOURNEL, B., ANGOT, S., JOYER, P., Decontamination of Phebus Experimental Target Chamber Using Sprayed Foam, Proc. 10th Int. Conf. Nucl. Eng. (ICONE 10), Arlington, VA, 14–18 April 2002 (2002) pp. 835–844.

[I–26] CHERNIKOV, M.A., UTROBIN, D.V., FELITSYN, M.A., Low-Waste Decontamination Technologies for Radioactively Contaminated Surfaces, Nuclear and Radiation Safety (2017), http://tvel2017.ru/nuclear-and-radiation-safety/en/

[I–27] HARMER, D.E., WHITE, J.L., "Results to date of the Dresden-1 chemical cleaning", Decontamination and Decommissioning of Nuclear Facilities (OSTERHOUT, M.M., Ed.) Springer, Boston, MA (1980).

[I–28] OECD NUCLEAR ENERGY AGENCY, Decontamination Methods as Related to Decommissioning of Nuclear Facilities: Report by an NEA Group of Experts, NEA, Paris (1981).

[I–29] ELECTRIC POWER RESEARCH INSTITUTE, Maine Yankee Decommissioning Experience Report — Detailed Experiences 1997–2004, EPRI, Washington, D.C. (2005), https://decommissioningcollaborative.org/wp-content/uploads/2020/07/my-epri-report-2005.pdf

[I–30] ENCAPSULATION TECHNOLOGY, Passive Aerosol Generator Model PK-2000 (Fogging Machine) (2013), http://www.srs.gov/general/enviro/rosc/encap.htm

[I–31] FARRELL, C., Modifying Remote Technology for Magnox Storage Pond, Nucl. Decom. Rep. 6 (2012) 16–19.

Annex II

HAZARDS DURING DECONTAMINATION PROCESSES

II–1. OVERVIEW

As discussed in Section 4.2.6, a range of potential hazards are to be considered in selecting a decontamination process. Some of these hazards are linked to the activity of the SSCs that require decontamination, or to the environment. Others are related to the physical conditions, to the movements of people and tools, or to the handling of removed materials and waste management. Identifying and anticipating hazards is key to an accurate health and safety assessment, and to the success of the decontamination campaign. Hazards refer to the potential for occupational injury or illness and related time loss, damage to property, or impacts on the public or the environment. In increasing order of impact, the results of an incident or accident during decontamination include:

— Near misses (no damage, but these can be indicators of unhealthy, unsafe trends or systematic deviations eventually leading to more serious events);
— Inconvenience (this may include worker discomfort and reduced productivity);
— Project delays;
— Reduced quality of work;
— Deterioration of equipment and surroundings;
— Minor injuries/first aid cases;
— Disabling injuries;
— Fatalities.

II–2. COMMON HAZARDS

Common hazards in nuclear decontamination can be categorized according to impacting substances, physical and chemical parameters, or operations (combinations of many hazards are quite common). A non-exhaustive list of hazards is given in Sections II–2.1–II–2.3.

II–2.1. Impacting substances

Radioactive (including radiation and contamination hazards), carcinogenic, allergenic, corrosive, pyrophoric, toxic, genetically impairing, explosive, non-ionizing radiation, biological, combustible, or asphyxiating (vapours and fumes) substances.

II–2.2. Physical and chemical parameters

High pressure, electricity, rough and irregular surfaces, slippery surfaces, sharp edges, hanging loads, hot surfaces, working at heights (ceilings, roofs or scaffolding), extremely hot or cold workplaces, steam, noise, confined work areas, inadequate ventilation, congested environments, PPE, or laser beams.

II–2.3. Operations

Production, handling (uploading and downloading), removal, transport or storage of materials, secondary waste, tools, consumables, hazardous substances and equipment, using cranes and hoists, chemical mixing, handling glassware and other fragile items, heating chemicals, high pressure water jetting and activating high voltage devices.

An industrial health and safety programme in decontamination may include the following elements:

— Monitor the workplace (including air, alpha, beta, gamma, temperature and humidity);
— Identify hazards (based on past activities);
— Consider changing conditions (increase of contamination levels) and reassess hazards accordingly;
— Control hazards (protect workers, look for unfavourable trends);
— Establish monitoring frequency (continuous, real time, periodic, random);
— Document relevant conditions and facts;
— Establish emergency procedures.

Table II–1 provides a non-exhaustive list of commonly encountered hazard types, their consequences and prevention/mitigation measures.

TABLE II–1. SUMMARY OF COMMONLY ENCOUNTERED HAZARDS

Type	Hazard	Consequences	Prevention/mitigation
Chemical/ radiological	Inhalation	Long term health impacts	Use of proper PPE when cleaning contaminated surfaces. Medical monitoring. Properly designed and maintained ventilation systems. Safe work procedures. Worker education and training
Chemical	Unplanned corrosion of metals	Damage to or destruction of materials or components, resulting in loss of functionality	Understanding of substrate material prior to undertaking activity. Substitution with less harmful product
Chemical/mechanical	Spread of contamination on-site and off-site	Worker contamination; time, costs and human resources wasted; possible impact on public and environment	Better techniques/procedures, barriers and timely cleanup
Chemical/ mechanical/electrical	Explosive/ Pressurized gases/ Combustible gas, liquid, solid, or aerosol	Damage to property, immediate health impacts	Substitution with less harmful product. Adequate ventilation. Proper storage to minimize fire and explosion hazards. Safe work procedures, including transportation. Worker education/training. Good housekeeping. PPE
Chemical	Pyrophoric liquid or solid	A pyrophoric substance is one that, even in small quantities, is liable to ignite after contact with air	Adequate air suppression. Proper storage to minimize fire and explosion hazards. Safe work procedures, including transportation. Worker education/training. Good housekeeping. PPE
Chemical	Toxic	Long term health impacts	Substitution with less harmful product. Properly designed and maintained ventilation systems. Safe work procedures. Worker education and training. PPE

TABLE II–1. SUMMARY OF COMMONLY ENCOUNTERED HAZARDS (cont.)

Type	Hazard	Consequences	Prevention/mitigation
Chemical	Allergenic	Respiratory or skin sensitization (long term health impacts)	Substitution with other product. PPE
Chemical	Corrosive	Skin burning or irritation (immediate medical attention, time loss)	Safe work procedures. Worker education/ training. PPE
Biological	Biological pathogens, bird droppings, rodents, insects, medical waste	Acute toxicity (immediate medical attention, time loss, illness, disease, death)	Compliance with all infection prevention and control practices. Immunization programme (where appropriate), worker education/ training, medical monitoring
Electrical	Electric shock, short circuit	Immediate medical attention, time loss, death	Safe work procedures, worker education/ training. PPE
Electrical	Loss of power	Safety related equipment failure, resulting in a range of possible incidents/accidents (impacts on workers, public, environment)	A planning issue, secondary backup power may be considered. Fail safety design and procedure
Ergonomics	Strains, sprains	Damage of tissue due to overexertion (possible medical attention, time loss)	Safe work procedures. Worker education/ training. Reduce manual handling wherever possible Arrange packaging and equipment in easy reach. Early reporting of symptoms of ergonomic concerns
Ergonomics	Human error	Lack of fail-safe functions results in a range of possible incidents/accidents	Safe work procedures. Worker education/ training. Improvements of working environment
Fall	Slip/trip	Situations resulting in falls from height or walking surfaces (slippery floors, poor housekeeping, rough surfaces, exposed ledges). Consequences may include medical attention, time loss, permanent disability, death	Perform regular maintenance on flooring, stairwells and handrails. Inspect ladders prior to use. Worker education/training. Maintain good housekeeping practices. Ensure adequate lighting. PPE, appropriate clothing/ footwear
Fire/heat	Burns	Safety related equipment failure resulting in a range of possible incidents/accidents (impacts to workers, public, environment)	Worker education/training. Safe work procedures. Access to firefighting equipment (blankets and extinguishers). PPE. Regular maintenance of fire detection/suppression systems. Regular practice of fire drills
Vibration	Fatigue	Vibration can cause damage to nerve endings or material fatigue, which in turn can result in a range of possible incidents/accidents	Replace equipment or processes. Instruct workers to recognize symptoms and alert supervisor. Improved use of auxiliary aids. Regular rotation of workers

TABLE II–1. SUMMARY OF COMMONLY ENCOUNTERED HAZARDS (cont.)

Type	Hazard	Consequences	Prevention/mitigation
Mechanical	Failure, structural collapse, unexpected leaks	A range of possible incidents/accidents (impacts on workers, public, environment)	Regular maintenance and testing of equipment and integrity of structures. Safe working and emergency procedures regularly updated and exercised
Mechanical	Caught by/caught in/caught between	Skin, muscle, or a body part exposed to crushing, cutting, tearing, shearing items or components. Consequences may include immediate medical attention, time loss, permanent disability, death	Worker education/training. Safe work procedures. Consideration of clothing/footwear near the working environment
Noise	Noisy equipment, operation environment, explosions	Hearing damage or inability to communicate. Consequences may include immediate medical attention, time loss, and a range of possible incidents/accidents	Substitution with quieter equipment or processes. Worker training. Audiometric testing. Assess noise levels and perform routine monitoring. Signs notifying of noisy areas. Hearing protection devices, minimize exposure time
Radiation	Overexposure to ionizing radiation (alpha, beta, gamma)	Tissue damage by cell ionization (immediate medical attention, time loss, restriction on further working in controlled areas, death)	A planning issue. Worker education/training. Sufficient ongoing monitoring. Radiation protection programme. Safe work procedures. PPE
Radiation	Non-ionizing radiation (UV, infrared, laser beams, microwaves)	Injury to tissues (immediate medical attention, time loss)	A planning issue. Worker education/training. Sufficient ongoing monitoring. Safe work procedures. PPE
Struck by	Falling tools and other objects	Occupational injury or death	Worker education. Safe work procedures, PPE
Struck against surfaces	Sharp edges, rough surfaces	Injury to a body part (medical attention, time loss)	Worker education. Safe work procedures, PPE
Temperature	Extreme heat/cold	Heat stress/stroke, exhaustion, dehydration or hyperthermia/hypothermia (immediate medical attention, time loss)	Portable ventilation devices. Worker education about the effects of exposure. Communication. Work–rest cycles. Work scheduling to avoid long periods of exposure. Provision of water. PPE
Visibility	Impaired	Lack of lighting or blocked vision can result in procedural errors, and ultimately in a range of possible incidents/accidents	Planning issue. Supplementary lighting may be required. Recognize and report unfavourable conditions

Annex III

NATIONAL CASE STUDIES

III–1. CASE STUDY 1: DECONTAMINATION OF CARTRIDGE COOLING PONDS, MAGNOX REACTOR, UNITED KINGDOM

The first generation Magnox reactor fleet comprised a unique set of power reactors designed and commissioned in the United Kingdom from the 1950s to the 1970s. The reactors were all different in their design but had unique commonality in the cooling of the irradiated fuel elements. Irradiated fuel was recovered from the reactors and placed underwater in a pond located adjacent to the reactor buildings, to allow the dissipation of heat before the material was deemed safe for transfer to Sellafield for reprocessing. Irradiated fuel was typically stored in the cooling ponds for a maximum of six months, after which time it would begin to degrade, making retrieval much more difficult. All Magnox reactors have now reached the end of their operational life and the majority of the Magnox reactors have been defuelled and are in an advanced stage of post-operational clean-out and decommissioning. They provide a useful case study to evaluate how different decontamination techniques can be applied to similar situations to achieve common goals.

III–1.1. Decontamination strategy

The first stage in designing a decontamination strategy for the cooling ponds was to ascertain the levels of contamination both in the residual cooling water and sorbed to the pond walls and floor, and then to further understand how far the contamination had penetrated the pond's concrete construction materials. Using both in situ and laboratory assessment techniques, a comprehensive characterization programme was instigated. This involved collecting a variety of pond water samples at different depths through the water column, sampling any sludge that had accumulated in the bottom of the pond, and both taking surface swab samples off the side of the pond wall and drilling and extracting a series of concrete cores from the pond walls for destructive analysis to ascertain how far contamination had penetrated.

The results from the analysis of these types of samples varied between the Magnox sites — a consequence of how each of the ponds was constructed, whether the walls were pretreated and whether the pond had been tanked/sealed prior to finishing with a paint/coating. It was also found that some of the ponds had been treated with chemicals through their lifespan and others had the water refreshed more frequently; this further complicated the decontamination strategy.

At the Hunterston A power station site, a comprehensive characterization process was designed and implemented to understand the depth of penetration and the distribution of the activity [III–1]. Using a combination of in situ and laboratory techniques, assessment of the samples suggested that the pond activity levels were typically 200 Bq/ml, attributed to Cs and Sr, with minor amounts of Pu and Am; however, the activity was heterogeneously distributed over the pond walls. Despite the years of operation, 95% of the activity was predicted to be contained within the pond coating, a result of the pigment used in the paint and, thus, removal of 4–6 mm of the pond surface would remove all the coating and the top couple of concrete, which would inadvertently remove 99% of the total activity.

Assessment of this characterization data initiated the development of a decontamination strategy. The strategy assessed the level and type of contamination present, accessibility, availability of technology and experience of workers, as well as time and cost implications. At some of the Magnox sites, the levels of contamination inferred from swabbing the pond surface prior to drawdown eliminated many decontamination techniques on the basis that workers would be exposed to dose limits >1 mSv/h, which exceeded their permitted working limits. Hence, in some cases this factor alone determined that more

FIG. III–1. Characterization of materials extracted from the cooling pond at Hunterston A power station, United Kingdom, prior to the development of a decontamination strategy. Reproduced with permission from the Hunterston A site, Magnox Ltd, United Kingdom.

aggressive decontamination techniques were needed, and special precautions were required in terms of managing radiation exposure (Fig. III–1).

At Hunterston A, it was recognized that the remaining cooling water provided a good bioshield from the more contaminated lower parts of the pond. Thus, instead of draining the pond, which had the potential to release airborne activity, it was proposed to pump the pond slowly whilst simultaneously instigating the selected decontamination technique. Plastic pontoons were lowered onto the surface of the pond to allow the workers to apply the decontamination method as close as possible to the pond walls. As the water level from the pond was reduced from drawdown, operators used UHP water jetting to remove the coated surface. The main reasons for the selection of this technique were that:

— It would produce no secondary effluent other than water, and this would then be contained within the pond and would not significantly add to the inventory of waste generated;
— It was efficient in terms of time and cost, considering the surface area requiring treatment;
— It was readily available;
— Sufficient individuals qualified and experienced in using the technology were available.

The pond walls were treated and then subsequent analysis was undertaken to ascertain whether the technique was successful. There was no trace of activity more than 10 mm into the core walls, even at locations where the pond coating had been physically damaged and, thus, the exercise was deemed successful.

However, a great deal of learning from experience was realized from this exercise, the use of pontoons in this manor was recognized and, thus, subsequent decontamination efforts at Bradwell and Hinkley Point A also followed a similar methodology. However, whilst UHP water jetting was selected at Hunterston, the technique generated excessive quantities of sludge, which then required further treatment and it was not effective on the floors of the pond (the moving effluent tended to recontaminate areas).

Hence, at Trawsfynydd a dry scabbling system was used to cut out up to 40 mm of the wall surface and at Chappelcross, because the pond floor was not flat, liquid nitrogen blasting was used to strip the contamination from the substrate. At Hinkley Point A, personnel have begun to dry shave a 10 mm layer of concrete off the reactor one spent fuel pond walls using a five disc rail cutter, which bolts on to the wall, cuts a 2 m × 2.3 m area and generates a powder waste with no wet sludge. At Bradwell, once the walls were decontaminated, a sealant was painted on to stabilize the surfaces in advance of the site entering a period of quiescence as care and maintenance progressed.

III–1.2. Conclusions

By understanding the behaviour of radionuclides and how they interact with the substrate of concern, it is possible to design and develop a sophisticated strategy that takes advantage of this understanding. The detailed analysis of the cores from Hunterston A showed that most contaminants in concrete are in the cement paste surrounding the rock aggregate. The rock aggregate, once separated from the contaminated paste (using concentrated nitric acid), was of such low activity level that it could be classified as 'out of scope, or free release' and, thus, could be disposed of at very little cost. Understanding a material at this level is relatively expensive, but it serves the purpose of realizing the implications of developing a comprehensive strategy.

III–2. CASE STUDY 2: DECONTAMINATION EXPERIENCE AT ÚJV ŘEŽ, CZECH REPUBLIC

III–2.1. Introduction

The ÚJV Řež, a. s. (ÚJV) was established in 1955 as the Nuclear Research Institute Řež. The main issues addressed at ÚJV in recent decades have included research, development and technical services for nuclear power plants, development of chemical technologies for the nuclear fuel cycle and irradiation services for research and development.

A significant amount of radioactive waste was generated by ÚJV's activities (operational waste and waste from reconstruction of nuclear facilities) [III–2]. Until the 1980s, disposal of radioactive waste in repositories was free of charge in the former Czechoslovakia and, hence, there was no need for decontamination. The easily processed waste was disposed of, and the difficult to process radioactive waste (mainly large items) was stored under inadequate conditions on the ÚJV site (Figs III–2 to III–4).

With the exception of an initially simple operational decontamination procedure, no further decontamination was carried out. In the 1990s, new, stricter legislation was adopted and the earlier practice was no longer accepted. Further, the disposal of radioactive waste was charged for, with a significant (constantly increasing) disposal fee being applied. After many years of activity, there were numerous obsolete nuclear facilities to be decommissioned. In 2003, decontamination and decommissioning started together with processing of accumulated legacy radioactive waste. The total amount of radioactive waste for processing in the first stage was approximately 950 m^3. It is estimated that, in the forthcoming second phase, approximately 500 m^3 of radioactive waste will require processing. The above mentioned facts meant that there was a necessity to perform decontamination to enable the unrestricted release of the material. A cost evaluation of the possible decontamination technologies was undertaken and compared with the estimated cost of direct disposal.

An evaluation of standard industrial decontamination technologies with small modifications was undertaken. Those techniques that were deemed cost effective and fit for purpose were purchased for use in this application (see Table III–1). Photographs of the workshop at which the segmentation and decontamination were undertaken are shown in Fig. III–5.

FIG. III–2. Bulk storage of radioactive waste outdoors. Courtesy of Nuclear Research Institute Řež, Czech Republic.

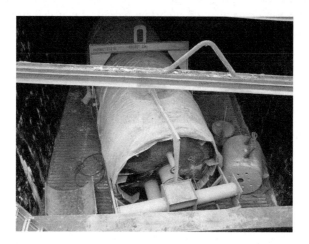

FIG. III–3. Storage of the old VVR-S research reactor vessel. Courtesy of Nuclear Research Institute Řež, Czech Republic.

FIG. III–4. Various stored radioactive wastes. Courtesy of Nuclear Research Institute Řež, Czech Republic.

TABLE III–1. LIST OF METHODS SELECTED FOR DECONTAMINATION

Techniques used for decontamination campaign
Vacuuming (vacuum cleaner with HEPA filter)
High pressure water jet
Chemical decontamination
Foam decontamination
Ultrasonic decontamination
Dry ice blasting
Grit blasting (in box)
Equipment for concrete decontamination

FIG. III–5. Photographs of the workshop where the segmentation and decontamination were undertaken Left: Segmentation and decontamination workshop, right: decontamination box. Courtesy of Nuclear Research Institute Řež, Czech Republic.

III–2.2. Applications of selected technologies

III–2.2.1. Pipe decontamination using a combination of techniques

A stainless steel pipe network with a total length of 410 m, situated in an underground concrete corridor, was used for the transfer of liquid radioactive waste. The total amount of contaminated metal was approximately 20 t.

A standard mechanical saw was used to cut the pipes. Some pipe parts (joints and flanges and corroded parts) were sent for conditioning without treatment. A high pressure water jet was used for internal and external preliminary decontamination of the pipes. An ultrasonic bath was used for internal decontamination of pipes (Fig. III–6). The decontamination was successful in most cases; however, some pipes required additional mechanical decontamination by a special single purpose instrument (an abrasive rotating device). After decontamination was performed, the residual contamination levels were assessed using a special tube detector. Any parts with residual contamination were cut out and disposed of. Approximately 90% of the pipes were free released.

III–2.2.2. Decontamination of storage tanks

Four cylindrical steel tanks (length 9.5 m, diameter 3 m, weight ~10 t), each with a capacity of 63 m³ located in underground bunkers, served for the receipt of liquid radioactive waste. The original aim of the project was to decontaminate and dismantle the tanks prior to the installation of new tanks. The tanks were decontaminated using high pressure water jetting, chemical foam and simple mechanical methods. After investigation of their integrity, decontamination was deemed so successful that the tanks could remain in use once a polyethylene or stainless steel lining had been installed (Fig. III–7).

III–2.2.3. Decontamination of a hot cell

A hot cell constructed from cast iron, used for dissolving irradiated fuel, was found to be contaminated. Previous decontamination efforts led to the coating of internal and external surfaces. During decontamination, the coating was partly removed by a paint remover at first and then dry ice blasting was applied. A small amount of secondary waste was generated. The process was successful, and the hot cell was reused.

FIG. III–6. Ultrasonic decontamination of pipes. Courtesy of Nuclear Research Institute Řež, Czech Republic.

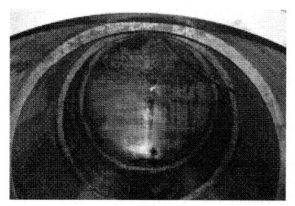 

(a) The tank after decontamination before installation of the liner   *(b) The tank after installation of the stainless steel lining*

FIG. III–7. Successful decontamination to extend the lifetime of storage tanks. Courtesy of Nuclear Research Institute Řež, Czech Republic.

 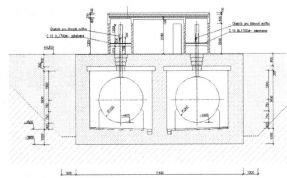

(a) Side-profile view of the bunker   (b) Plan of the bunker

FIG. III–8. Decay tanks used for storage and decay of concentrated short lived radioactive waste. Courtesy of Nuclear Research Institute Řež, Czech Republic.

III–2.2.4. Decay tanks

Decay tanks had been in use since 1961. The tanks were designed for storage and decay of concentrated short lived radioactive waste, but waste containing long lived radionuclides was also shipped there. The building is submerged on three sides (see Fig. III–8). It contains two cylindrical tanks (length 9.5 m, diameter 3 m, weight ~10 metric t), each with a capacity of 63 m^3. The decay tanks are made from structural steel jacketed by stainless steel inside the vessel. They are placed into two separate concrete bunkers located partially below ground with an auxiliary building above the bunkers.

The tanks did not solely contain liquid radioactive waste; one of them, tank B, also contained solid waste. The main identified radioisotope was ^{137}Cs. Solid radioactive waste consisted of tins with irradiated metallic samples and residues of spent fuel. The maximum dose rate (hundreds of mSv/h) was detected above the solid radioactive waste. The procedure for decontamination comprised the following steps:

(a) A structure above the decay tanks was constructed;
(b) The liquid waste was removed from the tanks;
(c) A special remote controlled manipulator was installed in the tank inlet to remove solid waste;
(d) This solid radioactive waste was removed and placed in a shielding container for transport to a hot cell facility for conditioning (the interior of the empty tank is shown in Fig. III–9);
(e) Access was required to allow future dismantling of the tanks;
(f) The tanks were then decontaminated by a high pressure water jet, chemical and mechanical decontamination methods (Fig. III–10);
(g) The tank surfaces were evaluated and hot spots removed (Fig. III–11).

The interior of tank B, after final decontamination, can be seen in Fig. III–12.

III–2.2.5. Gloveboxes

Gloveboxes were used for handling alpha radionuclides (U, Pu, Am, Np) in ÚJV Řež, a. s., Czech Republic (Fig. III–13). The boxes became obsolete, and it was decided to decommission them. The boxes were heavily contaminated by alpha radionuclides and their dismantlement represented a significant radiation risk. During dismantling of similar boxes several years ago, a significant amount of americium was released and staff were contaminated internally with it. Although this was caused by inadequate radiation protection measures, it demonstrated that contamination could be released during dismantling (even if the remaining contamination was fixed).

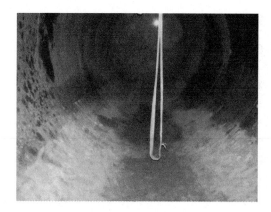

FIG. III–9. Interior of tank B after the removal of radioactive waste. Courtesy of Nuclear Research Institute Řež, Czech Republic.

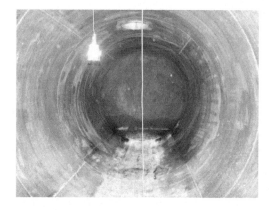

FIG. III–10. Interior of tank B after preliminary decontamination (water jetting decontamination). Courtesy of Nuclear Research Institute Řež, Czech Republic.

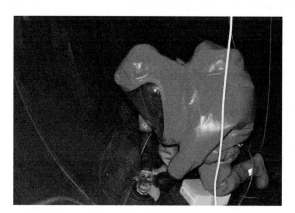

FIG. III–11. Removal of hot spots after water jetting decontamination. Courtesy of Nuclear Research Institute Řež, Czech Republic.

Based on this previous learning from experience, it was decided that the gloveboxes would not be further dismantled but disposed of as one piece in non-standard units after approval by the authority responsible for operation of the repository. The boxes were disconnected and sealed. Before that, the internal space of the boxes was filled with other radioactive waste to maximize the use of the boxes. The boxes were then inserted into the disposal units (Fig. III–14), grouted and sent for disposal (Fig. III–15).

FIG. III–12. Interior of tank B after final decontamination. Courtesy of Nuclear Research Institute Řež, Czech Republic.

FIG. III–13. Gloveboxes used for handling alpha radionuclides (U, Pu, Am, Np) in ÚJV Řež before decontamination for decommissioning. Courtesy of Nuclear Research Institute Řež, Czech Republic.

FIG. III–14. Inserting a glovebox used for handling alpha radionuclides (U, Pu, Am, Np) into non-standard disposal units. Courtesy of Nuclear Research Institute Řež, Czech Republic.

III–2.3. Conclusions

Assessment of the structure and components indicated that different decontamination techniques

FIG. III–15. Gloveboxes used for handling alpha radionuclides (U, Pu, Am, Np) stored in non-standard disposal units. Courtesy of Nuclear Research Institute Řež, Czech Republic.

need to be deployed to optimize performance. In some cases, multiple techniques were undertaken to achieve the desired outcome. In addition, in some cases small modifications were needed prior to deploying a particular technology. In other cases, it was concluded, on the basis of previous experience or safety considerations, that decontamination was not the best option.

REFERENCES TO ANNEX III

[III–1]   BOWER, W.R., et al., Characterising legacy spent nuclear fuel pond materials using microfocus X-ray absorption spectroscopy, J. Hazard. Mater. **317** (2016) 97–107.

[III–2]   PODLAHA, J., Remediation of Old Environmental Liabilities in the Nuclear Research Institute Řež plc, paper presented at 18th Annual Nuclear Sem. and Information Mtg, Častá-Papiernička, Slovakia, 2010.

Annex IV

EXAMPLE OF A DECONTAMINATION PROCEDURE

Following the establishment of a decontamination strategy, a series of working procedures can be developed to describe the working practices. This annex contains a typical example of a written procedure that may be used for small scale decontamination applications and for basic training in how to construct a procedure). An example of a cover page is illustrated in Fig. IV–1. The following subsections of this annex provide an example of the contents to be included in a written decontamination procedure.

IV–1. SCOPE

The procedure describes the working mode to be applied for the decontamination of surfaces/contaminated objects/on-site transport and manipulation equipment.

IV–2. DOMAIN

The procedure is applied in the (company, department, laboratory name) and establishes the steps to be followed for decontamination of surfaces/contaminated objects/on-site transport and manipulation equipment to be performed by the decontamination team established according to the work permit no. XXX.

IV–3. REFERENCE DOCUMENTS

Examples of reference documents that can be added are:

— Decontamination strategy plan;
— Quality management, code......., revision in force;
— Law no....... on.........;
— Fundamental norm no..... on.......;
— Working and maintenance instruction for equipment...1, 2, 3 (equipment used in the decontamination process, such as ultrasound bath, water jet devices, others)..........;
— Special instructions for substances that will be used in the decontamination process.

(Other types of reference documents need to be listed if relevant.).

IV–4. DEFINITIONS AND ACRONYMS

IV–4.1. Definitions

Listing of definitions, such as :

— Fixed contamination — ...;
— Unfixed contamination — ...;

— Ultrasound decontamination — ...;

(Any definition that is considered relevant in the scope of procedure)

Approved by.....................

WORKING PROCEDURE

Radioactive decontamination of working surfaces, contaminated objects, transport and manipulation equipment

(title)

Code: .... /Revision: ......

Applicable from: ......(date)......

Controlled copy Uncontrolled copy

Copy no:

VERIFIED BY:		VERIFIED BY:	
Name:		Name:	
Position:		Position:	
Signature:		Signature:	
ELABORATED BY:			
Name:			
Position:			
Signature:			

FIG. IV–1. Example of a cover page for a written decontamination procedure.

IV–4.2. Acronyms

List of acronyms:

— RWP — radiological work permit;
— QC — quality control;
— PPE — personal protective equipment;
— DF — decontamination factor.

IV–5. RESPONSIBILITIES

IV–5.1. General Director/Administration Board/Head of Department

5.1.1. Approves the procedure.
5.1.2. Is assuring the financial and technical resources for the process implementation.
5.1.3. ….

IV–5.2. Head of operation (decontamination manager)

5.2.1. Elaborates the present procedure and the decontamination plan, including the radiological survey map.
5.2.2. Assures the proper personnel training for those who will be deploying the decontamination technique.
5.2.3. Is responsible for assuring the necessary materials and equipment.
5.2.4. Is supervising and is accountable for the decontamination works, assesses the results, ….
5.2.5. ….

IV–5.3. Radiological safety officer (RSO)/health physicist

5.3.1. Provides training of the decontamination team from the radiological safety point of view.
5.3.2. Decides to stop operations if the activity levels or dose rates are higher than those estimated.
5.3.3. Is responsible for the adequacy of the PPE to be used, and further for measuring the PPE, decontamination (if needed) and/or reuse or treatment as radioactive waste.
5.3.4. Acts to delimitate the decontaminated area and to restrict access.
5.3.5. In cases of worker contamination, improper ventilation operation, or any other abnormal situations, establishes the working period of time (dose sharing) and the monitoring needs after process termination (e.g. whole body counters).
5.3.6. ….

IV–5.4. Head of characterization laboratory

5.4.1. Trains the workers for sampling.
5.4.2. Coordinates the in-time characterization of samples, results interpretation and certificate delivery.
5.4.3. ….

IV–5.5. Quality assurance manager

5.5.1. Performs the revision of the procedure in accordance with the outcomes from operational experience.

5.5.2. ....

IV–5.6. Dosimetrist

5.6.1. Performs the dosimetry measurements, prior to and at the end of the decontamination process.

5.6.2. Performs dosimetry measurements on PPE, tools, materials which are removed from the contaminated area, at the end of the working day or at any time that they are needed.

5.6.3. Performs samplings from the contaminated surfaces for spectrometric measurements.

5.6.4. Records the measured values in the dedicated registries or databases.

5.6.5. Marks the hot spots.

5.6.6. ....

IV–5.7. Decontamination personnel/team

5.7.1. Enters the contaminated area only with the established PPE.

5.7.2. Performs the decontamination operations according to the decontamination plan, specific procedures and instructions, under coordination and/or supervision of the decontamination manager.

5.7.3. Performs the collection and sorting of the secondary wastes resulting from decontamination for further management.

5.7.4. At the end of work, performs radiological monitoring under RSO surveillance.

5.7.5. ....

IV–6. PREREQUISITES

6.1. The decontamination team is complete and knows how the calibrated equipment devoted for decontamination operations has to be used, knows and can apply the emergency plan in case of an incident or accident, knows the radiation protection rules (e.g. keeping the PPE on for the whole time that the team is in the area during decontamination, smoking not being allowed, etc.) ....

6.2. The selected decontamination method for... was applied and results are recorded in....

6.3. ....

IV–7. WORK/JOB DESCRIPTION

This section depends on the situation (i.e. surface contamination and/or deep contamination) and the work to be performed, and the materials to be used, according to the plan.

IV–7.1. Introduction

The decontamination method needs to take into consideration:

— The characteristics of radionuclides to be removed, such as type, and physical and chemical form;
— The physicochemical characteristics of the material to be decontaminated;
— The level of decontamination that needs to be achieved;
— The future utilization of the decontaminated material/object/...;
— The area where the decontamination will be performed;
— The nature and quantity of and the treatment methods for secondary wastes;
— The radiation protection measures taken during the decontamination process;
— The cost–benefit analysis (which could also lead to no decontamination, just direct treatment and conditioning for disposal).

Depending on the method chosen, consider the following:

(a) If chemical decontamination is applied — chemical decontamination is used to remove radioactive particles from a surface and objects using chemical substances at different concentrations. The efficiency of the chemical decontamination process can be expressed by means of a decontamination factor, DF, which is usually higher if the method is more corrosive. The chemical decontamination methods used are:

   (i) Concentrated methods, which implies utilization of chemical solutions at concentrations >5%. These methods are applied on those surfaces that are not susceptible to corrosion.

   (ii) Diluted methods or with low concentration, which implies utilization of chemical solutions at concentrations <5%. These methods are less corrosive and the use of smaller quantities of chemical substances and the secondary wastes do not usually cause problems but require longer decontamination periods. The DF achieved is lower.

   (iii) Foam and gel methods, which are used for decontamination of floors, walls, in generally easily accessible planes surfaces, although the DF is often not very high.

   (iv) Ultrasound methods, which implies the use of ultrasound waves in combination with chemical solutions in which contaminated objects are immersed.

   (v) The chemical decontamination agents used are:
      — Acids (hydrochloric and nitric) in general for pickling;
      — Bases (hydroxides, phosphates, carbonates);
      — Detergents or surfactants;
      — Complexing agents (citric acid, oxalic acid and ethylenediaminetetraacetic acid (EDTA));
      — Organic solvents (benzene, carbon tetrachloride, ethyl alcohol, acetone, toluene) for degreasing.

If mechanical decontamination is applied — the mechanical decontamination technique usually removes some thickness/layers of the material and is used for external surfaces that are easily accessible and when the reducing of dimensions is not important. The thickness of the removed layer is dependent on the method and the contamination depth. Mechanical methods can include scarifying, manual sanding, milling with special equipment, water and steam jets, water jets and abrasion.

IV–7.2. Equipment and devices used in the decontamination process

A list of equipment and devices that will be necessary in the process, including PPE, is given below.

IV–7.3. Decontamination process

In the case of objects to be decontaminated:

— The objects may need to be transported towards the designed decontamination area (in other situations, it may be more applicable to undertake the decontamination procedure in situ).
— After the objects are loaded in the area, the handling equipment (e.g. forklift) is checked for contamination, and if it is contaminated, the measures that are foreseen in IV–7.4 are applied.
— The level of contamination is established using radiometric methods and/or smear tests, which will be performed using various analytical techniques.
— The decontamination method is established based on the material, the nature of the contaminant and the object dimensions.
— The selected decontamination method is applied.
— After decontamination is performed, a radiological characterization is undertaken. If the results are not in compliance with the proposed objective, the decontamination method can be reapplied or combined with other techniques to achieve the desired outcome if needed.

— When the DF is achieved, the secondary wastes are transferred for further management. At the end of the work, the whole area is radiologically characterized and cleaned.

In the case of surfaces to be decontaminated:

— After measuring the surfaces, the contaminated areas are marked.
— The decontamination method is established based on the level of contamination determined using radiometric methods and/or smear tests performed using laboratory analysis methods).
— The selected decontamination method is applied.
— After decontamination is performed, radiological characterization is undertaken. If the results are not in compliance with the proposed objective, the decontamination method can be reapplied or combined with other techniques to achieve the desired outcome if needed.
— When the DF is achieved, the secondary wastes are transferred for further management. At the end of the work, the whole area is radiologically characterized and cleaned.

IV–7.4. Decontamination of equipment contaminated during decontamination processes

During the decontamination processes, the transport and handling devices can themselves become contaminated. This is why when equipment leaves the designated decontamination area it has to be checked to avoid spreading contamination.

The steps are quite similar; once the decontamination process has been applied, and has been checked again, if the contamination persists the decontamination procedure is applied. Decontamination has been achieved when the measured values are in compliance with the radiological safety requirements.

In general, a single method or combined methods can be applied because decontamination methods cannot be 'standardized'.

IV–8. ACCEPTANCE CRITERIA

IV–8.1. Measured values after decontamination

The measured values after decontamination will not exceed 0.04 Bq/cm^2 for alpha and 0.4 Bq/cm^2 for beta radiation (or as established by national legislation).

IV–8.2. Report evaluation

This will be performed by the decontamination manager and radiological safety officer/health physicist, and will be approved by….

IV–9. FORMS/RECORDS/ANNEXES

Record retention details should be documented and specified.

IV–9.1. Records

— Bulletin for control of contamination, code: …;
— Certificate for activity measurements on samples, code….
— ….

IV–9.2. Forms

— Registries (e.g. PPE decontamination and/or laundry, materials used, radioactive wastes resulting, ….), code…;
— Revisions control list;
— ….

IV–9.3. Annexes

— Specific instructions for chemicals;
— Work protection measures for working at high heights;
— ….

IV–10. REVISIONS

The types and extent of expertise required to deploy a decontamination facility vary widely with the type of technology and the complexity of the process. In order to optimize the decontamination process, it will be necessary to review the training and qualification, radiological and chemistry monitoring programmes, and equipment operation at regular periods. An example of a review of a decontamination procedure is presented in Table IV–1.

TABLE IV–1. REVISION TABLE

No. of revisions	Scope of revision	Pages revised	Author of revision/date
0	Initial elaboration as a result of quality management system implementation according with……. standard. or Initial elaboration as a consequence of implementing the decontamination technology using … (ultrasound, abrasive,…)	All	Name………………… /09.2014
1	Revision as a consequence of decontamination technologies modification/ optimization (chemistry modification, equipment replacement with different characteristics, other)	Page ... to ... and page…	Name……………. /01.2016

GLOSSARY

The following list provides the definitions of relevant terms used in this publication. All definitions are included in the IAEA Nuclear Safety and Security Glossary 2022 (Interim) Edition[1], except those with the * symbol.

acceptance criteria. Specified bounds on the value of a *functional indicator* or *condition indicator* used to assess the ability of a *structure, system or component* to perform its *design* function.

accident. Any unintended *event*, including operating errors, equipment *failures* and other mishaps, the consequences or potential consequences of which are not negligible from the point of view of *protection and safety*.

arrangements (for operations). The integrated set of infrastructural elements necessary to provide the capability for performing a specified function or task required to carry out a specified operation.

assessment. The *process*, and the result, of analysing systematically and evaluating the hazards associated with *sources* and *practices*, and associated *protection* and *safety* measures.

authorization. The granting by a *regulatory body* or other governmental body of written permission for a *person or organization* (the *operator*) to conduct specified *activities*.

background. The *dose* or *dose rate* (or an observed measure related to the *dose* or *dose rate*) attributable to all *sources* other than the one(s) specified.

barrier. A physical obstruction that prevents or inhibits the movement of people, radionuclides or some other phenomenon (e.g. fire), or provides shielding against *radiation*.

characterization. Determination of the nature and *activity* of radionuclides present in a specified place.

clearance. Removal of *regulatory control* by the *regulatory body* from *radioactive material* or *radioactive* objects within notified or authorized *facilities and activities*.

clearance level. A value, established by a *regulatory body* and expressed in terms of *activity concentration*, at or below which *regulatory control* may be removed from a *source* of *radiation* within a notified or authorized *practice*.

containment. Methods or physical structures designed to prevent or *control* the *release* and the *dispersion* of *radioactive substances*.

contamination. *Radioactive substances* on surfaces, or within solids, liquids or gases (including the human body), where their presence is unintended or undesirable, or the *process* giving rise to their presence in such places.

fixed contamination. Contamination other than non-fixed contamination.[2]

---

[1] INTERNATIONAL ATOMIC ENERGY AGENCY, IAEA Nuclear Safety and Security Glossary, Non-serial Publications, IAEA, Vienna (2022).

[2] See INTERNATIONAL ATOMIC ENERGY AGENCY, Regulations for the Safe Transport of Radioactive Material, IAEA Safety Standards Series No. SSR-6 (Rev.1), IAEA, Vienna (2018).

non-fixed contamination. Contamination that can be removed from a surface during routine conditions of transport.[3]

corrosion*. Progressive surface dissolution of a material. A term generally used for metals. Corrosion can be uniform over the surface of the material or non-uniform through enhanced corrosion in stressed areas at physical discontinuities.

cost–benefit analysis. A systematic technical and economic evaluation of the positive effects (benefits) and negative effects (disbenefits, including monetary costs) of undertaking an action.

decontamination. The complete or partial removal of *contamination* by a deliberate physical, chemical or biological *process*.

decontamination, chemical*. The removal or reduction of radioactive contamination from surfaces by chemical processes.

decontamination factor. The ratio of the *activity* per unit area (or per unit mass or volume) before a particular *decontamination* technique is applied to the activity per unit area (or per unit mass or volume) after application of the technique.

difficult to detect (hard to detect, hard to measure) nuclide*. A radionuclide whose activity is difficult to measure directly from the outside of an SSC by non-destructive assay means.

dose rate. The *dose* per unit time.

graded approach. For a system of *control*, such as a regulatory system or a *safety system*, a *process* or method in which the stringency of the *control* measures and conditions to be applied is commensurate, to the extent practicable, with the likelihood and possible consequences of, and the level of *risk* associated with, a loss of *control*.

hazard. The potential for harm or other detriment, especially for *radiation risks*; a factor or condition that might operate against *safety*.

incident. Any unintended *event*, including operating errors, equipment *failures*, *initiating events*, *accident precursors*, *near misses* or other mishaps, or unauthorized act, malicious or non-malicious, the consequences or potential consequences of which are not negligible from the point of view of *protection and safety*.

(The term incident is used to describe events that are, in effect, minor accidents, that is, that are only distinguished from accidents in terms of having less severe consequences. However, unlike an accident, an incident can be caused intentionally*.)

inspection. An *examination*, observation, *surveillance*, measurement or test undertaken to assess *structures, systems and components* and materials, as well as operational *activities*, technical *processes*, organizational *processes*, *procedures* and personnel competence.

---

[3] See INTERNATIONAL ATOMIC ENERGY AGENCY, Regulations for the Safe Transport of Radioactive Material, IAEA Safety Standards Series No. SSR-6 (Rev.1), IAEA, Vienna (2018).

knowledge management. An integrated, systematic approach to identifying, managing and sharing an organization's knowledge and enabling groups of people to create new knowledge collectively to help in achieving the organization's objectives.

licence. A legal document issued by the *regulatory body* granting *authorization* to perform specified *activities* relating to a facility or activity.

licensee. The holder of a current licence. The licensee is the person or organization having overall responsibility for a *facility or activity*.

maintenance. The organized *activity*, both administrative and technical, of keeping *structures, systems and components* in good operating condition, including both preventive and corrective (or *repair*) aspects.

management system. A set of interrelated or interacting elements (*system*) for establishing policies and objectives and enabling the objectives to be achieved in an efficient and effective manner.

minimization (of waste). The *process* of reducing the amount and *activity* of *radioactive waste* to a level as low as reasonably achievable, at all stages from the *design* of a *facility or activity* to *decommissioning*, by reducing the amount of *waste* generated and by means such as *recycling* and *reuse*, and *treatment* to reduce its *activity*, with due consideration for *secondary* waste as well as primary *waste*.

monitoring. 1. The measurement of *dose, dose rate* or *activity* for reasons relating to the *assessment* or *control* of *exposure* to *radiation* or exposure due to *radioactive substances*, and the interpretation of the results.
2. Continuous or periodic measurement of radiological or other parameters or determination of the status of a *structure, system or component*.

occupational exposure. *Exposure* of *workers* incurred in the course of their work.

operator. Any *person* or *organization* applying for *authorization* or authorized and/or responsible for *safety* when undertaking *activities* or in relation to any *nuclear facilities* or *sources* of *ionizing radiation*.

optimization (of protection and safety). The *process* of determining what level of *protection and safety* would result in the magnitude of *individual doses*, the number of individuals (*workers* and *members of the public*) subject to *exposure* and the likelihood of *exposure* being *as low as reasonably achievable*, economic and social factors being taken into account (*ALARA*).

procedure. A series of specified actions conducted in a certain order or manner.

process. 1. A course of action or proceeding, especially a series of progressive stages in the manufacture of a product or some other *operation*.
2. A set of interrelated or interacting *activities* that transforms inputs into outputs.

radioactive waste. For legal and regulatory purposes, material for which no further use is foreseen that contains, or is contaminated with, radionuclides at *activity concentrations* greater than *clearance levels* as established by the *regulatory body*.

radioactive waste management. All administrative and operational *activities* involved in the handling, *pretreatment, treatment, conditioning, transport, storage* and *disposal* of *radioactive waste*.

regulatory body. An authority or a system of authorities designated by the government of a State as having legal authority for conducting the regulatory *process*, including issuing *authorizations*, and thereby regulating the *nuclear, radiation, radioactive waste* and *transport safety*.

release. The action or process of setting free or being set free, or of allowing or being allowed to move or flow freely.

remediation. Any measures that may be carried out to reduce the *radiation exposure* due to existing *contamination* of land areas through actions applied to the *contamination* itself (the *source*) or to the *exposure pathways* to humans.

risk. A multiattribute quantity expressing *hazard*, danger or chance of harmful or injurious consequences associated with *exposures* or *potential exposures*. It relates to quantities such as the probability that specific deleterious consequences may arise and the magnitude and character of such consequences.

safe enclosure (during decommissioning)*[4]. A condition of a nuclear facility during the decommissioning process in which only surveillance and maintenance of the facility take place.

safety case. A collection of arguments and evidence in support of the *safety* of a *facility* or *activity*.

scabbling*[6]. Mechanical process of removing a thin layer of concrete from a structure. A typical scabbler uses several heads, each with several carbide or steel tips that peck at the concrete. It operates by pounding a number of tipped rods down onto the concrete surface in rapid succession. It may take several passes with the machine to achieve the desired depth.

secondary waste. Radioactive waste resulting as a byproduct from the processing of primary *radioactive waste*.

segregation (part of radioactive waste management definition). An activity where types of waste or material (radioactive or exempt) are separated or are kept separate on the basis of radiological, chemical and/or physical properties, to facilitate waste handling and/or processing.

service life. The period from initial *operation* to final withdrawal from service of a *structure, system or component*.

[stakeholder] (interested party). A person, company, etc., with a concern or interest in the activities and performance of an organization, business, system, etc.

storage. The holding of *radioactive sources, radioactive material, spent fuel* or *radioactive waste* in a *facility* that provides for their/its *containment*, with the intention of retrieval.

structures, systems and components (SSCs). A general term encompassing all of the elements (items) of a *facility* or *activity* that contribute to *protection and safety*, except human factors.

component. One of the parts that make up a *system*.

---

[4] Definition taken from INTERNATIONAL ATOMIC ENERGY AGENCY, Decommissioning of Pools in Nuclear Facilities, IAEA Nuclear Energy Series No. NW-T-2.6, IAEA, Vienna (2015).

structure. A passive element (e.g. buildings, vessels, shielding).

system. A set of *components* which interact according to a *design* so as to perform a specific (active) function, in which an element of the *system* can be another *system*, called a subsystem.

technology readiness assessment*. A systematic, metrics based process and accompanying report that assesses the maturity of certain technologies used in systems.

technology readiness levels*. A type of measurement system used to assess the maturity level of a particular technology. Each technology project is evaluated against the parameters for each technology level and is then assigned a technology readiness level rating based on the project's progress.

unrestricted use. The use of an *area* or of material without any radiologically based restrictions.

waste. Material for which no further use is foreseen.

waste acceptance criteria. Quantitative or qualitative criteria specified by the *regulatory body*, or specified by an *operator* and approved by the *regulatory body*, for the *waste form* and *waste package* to be accepted by the *operator* of a *waste management facility*.

waste container. The vessel into which the *waste form* is placed for handling, *transport*, *storage* and/or eventual *disposal*; also the outer *barrier* protecting the *waste* from external intrusions. The *waste container* is a *component* of the *waste package*. For example, molten *high level waste* glass would be poured into a specially designed *container* (*canister*), where it would cool and solidify

CONTRIBUTORS TO DRAFTING AND REVIEW

Delabre, H.	International Atomic Energy Agency
Dragolici, F.	International Atomic Energy Agency
Gordon, I.A.	International Atomic Energy Agency
Laraia, M.	Consultant, Italy
Lust, M.	International Atomic Energy Agency
Meyer, W.	International Atomic Energy Agency
Nokhamzon, J.-G.	Consultant, France
Podlaha, J.	Nuclear Research Institute Řež, Czech Republic
Thompson, O.	National Nuclear Laboratory, United Kingdom

Consultants Meetings

Vienna, Austria: 11–15 December 2017; 25 May–1 June 2018; 17–21 February 2020

Structure of the IAEA Nuclear Energy Series*

Nuclear Energy Basic Principles
NE-BP

Nuclear Energy General Objectives
NG-O

1. Management Systems
NG-G-1.#
NG-T-1.#

2. Human Resources
NG-G-2.#
NG-T-2.#

3. Nuclear Infrastructure and Planning
NG-G-3.#
NG-T-3.#

4. Economics and Energy System Analysis
NG-G-4.#
NG-T-4.#

5. Stakeholder Involvement
NG-G-5.#
NG-T-5.#

6. Knowledge Management
NG-G-6.#
NG-T-6.#

Nuclear Reactor** Objectives
NR-O

1. Technology Development
NR-G-1.#
NR-T-1.#

2. Design, Construction and Commissioning of Nuclear Power Plants
NR-G-2.#
NR-T-2.#

3. Operation of Nuclear Power Plants
NR-G-3.#
NR-T-3.#

4. Non Electrical Applications
NR-G-4.#
NR-T-4.#

5. Research Reactors
NR-G-5.#
NR-T-5.#

Nuclear Fuel Cycle Objectives
NF-O

1. Exploration and Production of Raw Materials for Nuclear Energy
NF-G-1.#
NF-T-1.#

2. Fuel Engineering and Performance
NF-G-2.#
NF-T-2.#

3. Spent Fuel Management
NF-G-3.#
NF-T-3.#

4. Fuel Cycle Options
NF-G-4.#
NF-T-4.#

5. Nuclear Fuel Cycle Facilities
NF-G-5.#
NF-T-5.#

Radioactive Waste Management and Decommissioning Objectives
NW-O

1. Radioactive Waste Management
NW-G-1.#
NW-T-1.#

2. Decommissioning of Nuclear Facilities
NW-G-2.#
NW-T-2.#

3. Environmental Remediation
NW-G-3.#
NW-T-3.#

(*) as of 1 January 2020
(**) Formerly 'Nuclear Power' (NP)

Key
BP: Basic Principles
O: Objectives
G: Guides and Methodologies
T: Technical Reports
Nos 1–6: Topic designations
#: Guide or Report number

Examples
NG-G-3.1: Nuclear Energy General (NG), Guides and Methodologies (G),
Nuclear Infrastructure and Planning (topic 3), #1
NR-T-5.4: Nuclear Reactors (NR), Technical Report (T), Research Reactors (topic 5), #4
NF-T-3.6: Nuclear Fuel (NF), Technical Report (T), Spent Fuel Management (topic 3), #6
NW-G-1.1: Radioactive Waste Management and Decommissioning (NW), Guides
and Methodologies (G), Radioactive Waste Management (topic 1) #1

ORDERING LOCALLY

IAEA priced publications may be purchased from the sources listed below or from major local booksellers.

Orders for unpriced publications should be made directly to the IAEA. The contact details are given at the end of this list.

NORTH AMERICA

Bernan / Rowman & Littlefield

15250 NBN Way, Blue Ridge Summit, PA 17214, USA
Telephone: +1 800 462 6420 • Fax: +1 800 338 4550

Email: orders@rowman.com • Web site: www.rowman.com/bernan

REST OF WORLD

Please contact your preferred local supplier, or our lead distributor:

Eurospan Group

Gray's Inn House
127 Clerkenwell Road
London EC1R 5DB
United Kingdom

Trade orders and enquiries:

Telephone: +44 (0)176 760 4972 • Fax: +44 (0)176 760 1640
Email: eurospan@turpin-distribution.com

Individual orders:

www.eurospanbookstore.com/iaea

For further information:

Telephone: +44 (0)207 240 0856 • Fax: +44 (0)207 379 0609
Email: info@eurospangroup.com • Web site: www.eurospangroup.com

Orders for both priced and unpriced publications may be addressed directly to:

Marketing and Sales Unit
International Atomic Energy Agency
Vienna International Centre, PO Box 100, 1400 Vienna, Austria
Telephone: +43 1 2600 22529 or 22530 • Fax: +43 1 26007 22529
Email: sales.publications@iaea.org • Web site: www.iaea.org/publications